LMS & LNER STEAM LOCOMOTIVES

The Post War Era

Front Cover: **Fowler LMS 'Class 2P'** locomotive number 40630 passing Pontefract Monkhill Goods Yard in West Yorkshire with the 2pm Wakefield Kirkgate to Goole local 'Saturday-only' passenger service in February 1957. Pontefract Monkhill Station can be seen in the background.

The 4-4-0 locomotive was based at Normanton Shed and only worked this service due to it being a Saturday-only train. This was a rare treat for enthusiasts as Normanton-based engines seldom ventured onto this particular line.

Back Cover 1: **LNER 'Class K3'** locomotive number 61935 working hard as she ascends Gunness Bank near Scunthorpe in Lincolnshire with an express passenger train service to Cleethorpes in the mid-1950s.

Back Cover 2: **This photograph shows** LNER 'Class A4' 'Pacific' locomotive number 60031 *Golden Plover* in pristine condition, having just left the paint shop at Doncaster Works on 3 June 1948. She is turned out in the new British Railways white-lined, 'Express Blue' livery.

LMS & LNER
STEAM LOCOMOTIVES

The Post War Era

MALCOLM CLEGG

PEN & SWORD
TRANSPORT

AN IMPRINT OF PEN & SWORD BOOKS LTD.
YORKSHIRE – PHILADELPHIA

First published in Great Britain in 2021 by
Pen and Sword Transport
An imprint of
Pen & Sword Books Ltd.
Yorkshire - Philadelphia

Copyright © Malcolm Clegg, 2021

ISBN 978 1 52677 860 4

The right of Malcolm Clegg to be identified as author of this work has been asserted by him in accordance with the Copyright, Designs and Patents Act 1988.

A CIP catalogue record for this book is available from the British Library.

All rights reserved. No part of this book may be reproduced or transmitted in any form or by any means, electronic or mechanical including photocopying, recording or by any information storage and retrieval system, without permission from the Publisher in writing.

Typeset by SJmagic DESIGN SERVICES, India.
Printed and bound in India by Replika Press Pvt. Ltd.

Pen & Sword Books Ltd incorporates the Imprints of Pen & Sword Books Archaeology, Atlas, Aviation, Battleground, Discovery, Family History, History, Maritime, Military, Naval, Politics, Railways, Select, Transport, True Crime, Fiction, Frontline Books, Leo Cooper, Praetorian Press, Seaforth Publishing, Wharncliffe and White Owl.

For a complete list of Pen & Sword titles please contact

PEN & SWORD BOOKS LIMITED
47 Church Street, Barnsley, South Yorkshire, S70 2AS, England
E-mail: enquiries@pen-and-sword.co.uk
Website: www.pen-and-sword.co.uk

or

PEN AND SWORD BOOKS
1950 Lawrence Rd, Havertown, PA 19083, USA
E-mail: Uspen-and-sword@casematepublishers.com
Website: www.penandswordbooks.com

CONTENTS

Acknowledgement 6

Introduction 7

List of Abbreviations 10

Photographs and Captions 11–110

Locomotive Technical Details and General Information 111

Details of Locomotives Featured in this Book 112

    London, Midland and Scottish Railway (LMS) Locomotives 112–134

    London and North Eastern Railway (LNER) Locomotives 135–165

About the Author 166

ACKNOWLEDGEMENT

I would like to offer my sincere thanks and appreciation to Mr Peter Cookson, a retired schoolmaster, author and railway historian for providing the photographs and caption information for this book and allowing his photographs to be published. Without his contribution, this book could not have been written.

Copyright
All the photographs in this book are from the private collection of Mr Peter Cookson, © Peter Cookson.

INTRODUCTION

Rapid improvements in steam technology during the early nineteenth century led to the invention of the modern-day steam locomotive, which in turn gave birth to the modern-day railway network in Britain.

The opening of the Stockton and Darlington Railway in 1825 and the Liverpool and Manchester Railway in 1830, set the scene for a nationwide transport network which fuelled the industrial revolution and enabled goods and passengers to be transported speedily throughout Britain for the first time.

Prior to the railways being built, goods were transported on poorly maintained roads by horse drawn waggons or by barges on the expanding canal network which had been introduced just a few decades earlier and people seldom travelled more than a few miles from the towns or villages where they were born.

During the 1830s however, thousands of navigators (commonly referred to as navvies) were commissioned to construct railways for the many private railway companies that were being created at the time. The 1840s were witness to 'railway mania', as speculative frenzy saw more and more money being poured into the railways by people trying to make their fortunes. Some did make fortunes but many lost their investments and life savings to greedy and corrupt individuals. Fraud amongst railway speculators was widespread.

In 1846 alone, over 270 Acts of Parliament were passed, each one authorising the building of a new railway, and navvies continued to be employed in vast numbers to build them. Many of these navvies were from Ireland, having moved to England, Scotland and Wales to escape the extreme poverty caused in part by the effects of the potato famine. A considerable number brought their wives and families with them and they resided in the many 'shanty towns' which were being constructed alongside the railways, particularly in places where lengthy construction work on major projects such as viaducts and tunnels was being carried out.

Railway construction continued throughout the nineteenth century and into the twentieth. As more and more railways were being built, some railway companies amalgamated with others and many of the smaller railway companies were absorbed into larger ones. This led to the constant fluctuation in the numbers of railway companies operating in Britain at that time.

By 1921 however, with hundreds of mergers and amalgamations having already taken place, there were still in the region of 120 different railway companies operating in Britain. Many were losing money, having been adversely affected by the First World War, during which time the railways had been operating under government control.

As a result, the Railways Act of 1921 (commonly known as the Grouping Act) was introduced in an attempt to stem the losses by steering the numerous companies away from the fierce competition and rivalry with each other, by means of amalgamating them into just four large private companies which later became known as the 'big four'. Nationalisation of the railways was considered but rejected, partly on the grounds that nationalisation was likely to result in political interference and influence as well as poor management.

The provisions of the 1921 Railways Act came into effect on 1 January 1923 when the 'big four' railway companies were officially created. They were:

The Southern Railway Company (SR)
The Great Western Railway Company (GWR)
The London, Midland & Scottish Railway Company (LMS)
The London & North Eastern Railway Company (LNER)

These 'big four' railway companies operated until the railways were again taken under government control shortly after the outbreak of the Second World War in 1939.

At the end of that conflict, the railways were once again in serious financial difficulties and on the verge of bankruptcy. This resulted in further government intervention, which culminated in a decision to fully nationalise the railways. Nationalisation of Britain's railways came into effect on 1 January 1948 and the 'big four' railway companies were abolished. The name 'British Railways' was given to the network as a whole, which was operated by the British Transport Commission (BTC) until 1962.

In 1963, 'British Railways' became an independent statutory corporation, the 'British Railways Board', who operated the railway network until it was privatised in the 1990s (the Corporation traded as 'British Rail' from 1965 until privatisation took place).

This book and it's photographs, relate to steam locomotives which were designed and built for use on the two biggest of the 'big four' railway companies which operated between 1923 and 1948, namely the LMS and the LNER. Set out below, are some brief details about both companies.

The London, Midland & Scottish Railway (1923–1948)

The LMS Railway was the largest of the 'big four' railway companies and comprised the following major constituent railway companies that had existed prior to 1923:

The Caledonian Railway Company
The Furness Railway Company
The Glasgow and South Western Railway Company
The Highland Railway Company
The London and North Western Railway Company
The Midland Railway Company
The North Staffordshire Railway Company

The seven constituent companies between them had collectively acquired a vast number of other railway companies over the years, the last acquisition being the impressive Lancashire and Yorkshire Railway Company, who merged with the London and North Western Railway Company in 1922. Prior to their merger with the LNWR, the Lancashire and Yorkshire Railway alone had over 1,600 locomotives and the largest shipping fleet of all the pre-grouping railway companies and their merger with the London and North Western Railway Company meant that the LNWR itself had acquired almost 100 different railway companies since its foundation in 1846.

In addition to the seven constituent companies, the LMS inherited twenty-four subsidiary railway companies, and was the only one of the 'big four' companies to operate in England, Scotland, Wales and Northern Ireland. A large number of joint railway companies also became a part of the LMS Railway. The total route mileage inherited by the LMS Railway in 1923 was almost 7,800 miles (12,553 km).

When the LMS Railway Company was founded, it was not only the largest of the 'big four' Railway Companies, but it inherited the greatest number of locomotives. It acquired approximately 10,400 locomotives (excluding Northern Ireland) made up of over 400 different classes. The sheer variety and number of locomotives was a logistical nightmare when it came to planning their operation, distribution and maintenance. Programmes were quickly put in place to reduce the numbers and different classes of locomotive. Many old and inefficient locomotives were quickly withdrawn from service, together with others which had been inherited from many of the smaller railway companies.

Of the locomotive workshops inherited, the LMS decided to use four principal workshops, which were located at Crewe, Derby, Horwich (near Bolton in Lancashire) and St Rollox in Glasgow. These workshops became the mainstay of the LMS Railway for the building of new locomotives and the general servicing and maintenance of their working fleet.

The London & North Eastern Railway (1923–1948)

The LNER Company was the second largest of the 'big four' railway companies, and like the new LMS Company, it also comprised seven major constituent railway companies which had existed prior to 1923. These companies are listed as follows:

- Great Central Railway
- Great Eastern Railway
- Great Northern Railway
- Great North of Scotland Railway
- North British Railway
- North Eastern Railway
- Hull and Barnsley Railway
  (The NER and the H&B amalgamated in April 1922),

The LNER also owned twenty-six subsidiary railway companies as well as a number of joint railway companies. Over 7,700 steam locomotives were inherited by the LNER and the total route mileage of the LNER in 1923 was almost 6,600 miles (10,622 km).

Steam locomotives of both the LMS and LNER are the focal point of this book. Whilst many former LMS and LNER locomotives of various classes were never captured on film, a considerable number were photographed by railway enthusiasts and photographers over the years and many have been preserved for posterity.

This book depicts a number of LMS and LNER locomotives which were captured on film during the immediate post war period in the late 1940s and continuing throughout the 1950s. The photographs, which total 200 in number, show a variety of locomotives used by the former LMS and LNER railway companies. Many of these locomotives continued to operate on British Railways after nationalisation in 1948, and some even survived into the 1960s when steam traction was discontinued in favour of diesel and electric locomotives. Fortunately, a considerable number of steam locomotives, including former LMS and LNER locomotives, have been preserved and can be seen working on the numerous private heritage railways throughout Britain, as well as being displayed in museums and other institutions.

LIST OF ABBREVIATIONS

BR = British Railways
BTC = British Transport Commission
CR = Caledonian Railway
GCR = Great Central Railway
GER = Great Eastern Railway
GNR = Great Northern Railway
GNSR = Great North of Scotland Railway
HR = Highland Railway
H&B = Hull & Barnsley (Railway)
LMS = London, Midland and Scottish Railway
LNER = London & North Eastern Railway
LNWR = London and North Western Railway
LYR = Lancashire and Yorkshire Railway
MR = Midland Railway
NBR = North British Railway
NER = North Eastern Railway
NLR = North London Railway
ROD = Railway Operating Division (British Army, Royal Engineers)
SLS = Stephenson Locomotive Society (Rail tours)
WD = War Department (British Government)

LMS & LNER STEAM LOCOMOTIVES: THE POST WAR ERA

LMS 'Jubilee Class' locomotive number 45566 gathers speed as she passes Armley Road in Leeds, heading on a long journey to Scotland with a Leeds to Glasgow express passenger train in the mid-1950s.

This picture shows LNER 'Class K3' locomotive number 61868 working effortlessly through the rural countryside of West Yorkshire whilst travelling south towards Sheffield with an empty stock train in the summer of 1959.

LMS 3-cylinder compound number 41159 is seen here working hard as she makes her way through Heaton Norris, Stockport, with an express passenger train in the mid-1950s. Heaton Norris railway station closed in 1959.

Another photograph of number 41159, this time taken on 8 August 1953. In this picture she is working the 10.15am Manchester London Road to Bournemouth, '*Pines Express*' passenger train service near Cheadle Hulme, Manchester. The train travelled via Crewe, Birmingham and Bath Green Park Station (now closed), before using the former Somerset and Dorset railway line to Bournemouth, which closed in the 1960s. Manchester London Road Station was re-named Manchester Piccadilly in 1960.

A wonderful shot of this LNER 'Class K3' locomotive, number 61935, working hard as she ascends Gunness Bank near Scunthorpe in Lincolnshire whilst conveying passengers to the seaside resort of Cleethorpes in the mid-1950s.

The power of a 'Class A4' locomotive is quite evident as LNER number 60002 *Sir Murrough Wilson* hauls an up express passenger train through King Edward Bridge Junction, Newcastle, in the mid-1950s.

This Fairburn 'Class 4P' tank engine, number 2219, in pristine condition, was proudly displaying her LMS livery when caught on camera outside Crewe Works on 4 April 1948, not long after nationalisation of the railways. She was a Tilbury based locomotive at the time.

Another engine in immaculate condition is this LNER 'Class A8' tank engine, based at Middlesbrough but pictured at Darlington on 25 June 1949. She has just received her new British Railways Black livery and BR number 69860. The name 'BRITISH RAILWAYS' can be seen painted on the side of the locomotive in capital letters. The full name was only used on locomotives painted during the 1948/49 period, as the BR 'Cycling Lion' emblem replaced it in 1950.

Fowler LMS 'Class 2P' locomotive number 40630 is pictured passing Pontefract Monkhill Goods Yard in West Yorkshire with the 2pm Wakefield Kirkgate to Goole local 'Saturday-only' passenger service in February 1957. The scene is almost in silhouette due to the strong low sunshine, which, in conjunction with the plumes of smoke, creates a spectacular photograph. The platforms and footbridge of Pontefract Monkhill Station are clearly visible in the background of the picture.

Glasgow Queen Street Station in the mid-1950s is the scene of this picture, as LNER 'Class A4' locomotive number 60009 *Union of South Africa* makes her last-minute preparations before departing with an express passenger train. The fireman can be seen working hard on top of the tender, furiously raking the coal forward.

A rare 'Class V4' locomotive, number 61700 *Bantam Cock*, can be seen entering the picture from the left as she departs the station. Only two of these mixed traffic locomotives were ever built and they were the last locomotives designed by Nigel Gresley before he died in 1941. Both worked passenger services on the West Highland Line between 1943 and 1949.

LMS 'Fowler' 2-6-4 4P tank engine number 42407 scurrying along between stations with a three-coach local stopping train from Goole to Bradford Exchange via Wakefield Kirkgate in April 1957.

The picturesque setting of Brockholes Station near Huddersfield is the scene of another 2-6-4, 'Class 4P' tank locomotive. This time the 'Stanier' tank engine, number 42650, prepares to depart bunker first with a Penistone to Bradford Exchange passenger service (via Huddersfield) in 1958. The signal at the end of the station platform is an old former Lancashire and Yorkshire Railway semaphore signal, the remnant of a bygone era.

LNER (GNR), 'Class N1' Ivatt 0-6-2 tank engine number 69452 sits quietly in the platform at Drighlington and Adwalton Station near Bradford, West Yorkshire in May 1952. She is working a three-coach passenger train (cab first).

The positioning of the white lamp on the locomotive (single lamp/top/centre), denotes the train to be an ordinary (stopping) passenger service. The former Leeds, Bradford and Halifax Junction Railway line closed to all traffic in 1962, as did the railway stations.

A lone LNER (GNR) 'Class N1' tank engine, displaying the number E 9446, is pictured at Leeds in May 1948. Her former LNER number 9446, together with the prefix letter 'E', was a temporary number, applied during a three-month period only, following nationalisation of the railways on 1 January 1948. The prefix letters were used between January and March of that year whilst the new BR locomotive re-numbering system was being finalised. Engines that were given prefix letters to their old numbers retained them for a short period of time until they could be re-painted with their new BR numbers, in this case, 69446. Leeds Central Station, where this photograph was taken, was closed in April 1967.

'Black Five' locomotive number 44854 leaves Pontefract Baghill Railway Station with the 7.20am York to Sheffield ordinary passenger train on 11 April 1959. The plumes of grey and white smoke from the locomotive enhance this fine picture.

Built just three weeks before the railways were nationalised in 1948, this LNER 'Class B1' locomotive, number 61038 *Blacktail,* is preparing to depart Hull Paragon Station with an express passenger train to Leeds in the mid-1950s.

At first glance, this may seem like a freight train because of the vans behind the tender. However, passenger coaches (out of view) are attached to the three box vans which form part of the train and the two lamps positioned above the buffers on the front of the locomotive indicate it to be an express passenger train.

This small LMS 'Class 2F' engine, number 58198, makes her presence known as she trundles through Pontefract towards her destination at Rotherham Masborough in South Yorkshire with a slow mineral train on 2 February 1957. Built for the Midland Railway in 1880 she was well into her twilight years when this picture was taken but soldiered on for a couple more years until 1959, when she was withdrawn from service for scrap.

A typical picture of Leeds City Station as it was in the 1950s. There were two sides to the station. On the left can be seen what remains of the old Midland Railway Terminus Station of Leeds Wellington, built in 1846. A large part of that station (out of view) has been demolished and is now a car park.

The right of the picture, at a slightly higher level, is Leeds New Station which was a through station, built adjacent to Wellington Station in 1869, jointly by the LNWR and the NER Companies. In 1938, Wellington Station and New Station were combined to form Leeds City Station and whilst some modernisation has been carried out since this photograph was taken, the whole site is now occupied by the current Leeds Station, still often referred to as Leeds City Station. Four LMS locomotives appear in the photograph.

Another railway station, a short distance from this site, was the terminus of Leeds Central Station, built in 1854. Leeds Central, a joint GNR, NER and LYR Station, which played an important role in the railway history of Leeds, was closed in 1967.

This photograph, taken shortly after nationalisation, shows an un-named 'Class B1', number 1179, in LNER livery, working a Manchester to Sheffield express through Hazlehead, South Yorkshire on 25 March 1948. The engine was assigned to Sheffield Darnall Shed at the time.

Another LNER 'Class B1', number 61247 *Lord Burghley*, pictured as a light engine adjacent to St James' Bridge Station, Doncaster in 1955. She was based at Doncaster Shed when this picture was taken.

LMS 'Class 8F' heavy freight locomotive number 48202, passing under Sourgate Lane Bridge, Knottingley, West Yorkshire, with an up coal train during the summer of 1956. A person would not need to venture far in steam days before seeing a coal train, particularly in the coalfield areas of Britain.

It was a cold, autumn day in October 1956 when this photograph was taken, showing another 'Class 8F' locomotive, number 48537, fitted with a snow plough, taking on water at the Prince of Wales Colliery, Pontefract, West Yorkshire. At the same time, the driver, unaware he had been captured on film, was relieving himself of his own water. The POW colliery closed in 2002, with the loss of 500 jobs.

This rare photograph shows LNER, 2-6-0 'Class K2/1' number 61720 and an unidentified 'Class K3' locomotive, pictured at Grimsby Docks circa 1955/56, about to depart to an unknown destination whilst double-heading an express fish train.

The former GNR engine, 61720, was the first of just ten GNR 'Class H2' locomotives, built by Nigel Gresley in 1912 and later classified as LNER 'Class K1' locomotives. All ten engines were later rebuilt into LNER 'Class K2' locomotives and were re-classified as LNER 'Class K2/1' engines to distinguish them from brand new 'Class K2' engines which had also been built. These 'Class K1' locomotives should not be confused with the Thompson/Peppercorn 'Class K1' engines built between 1949 and 1950.

Another rare photograph of a former Gresley design locomotive shows this LNER 'Class K5' locomotive, number 61863, pictured outside Doncaster Works in 1950. She was originally built in 1925 as an LNER 'Class K3' 3-cylinder, mixed traffic locomotive. A total of 193 of these powerful mixed traffic locomotives were built between 1920 and 1937.

This locomotive became a unique member of the class when she was selected by Edward Thompson to be converted into a 2-cylinder locomotive in 1945 for ease of maintenance. After being rebuilt with two cylinders and a new boiler, she was classified as an LNER 'Class K5' locomotive. After the rebuild she continued to work successfully until 1960 when she retired from service for scrap. She was the only 'Class K5' engine ever built.

An unmistakable LMS 2-6-6-2T 'Beyer-Garratt' locomotive number 47977, pictured at Crewe Works in 1950. She had recently left the paint shop, having been painted in her new shiny black BR livery with the BR 'Lion on Wheel' logo on the cab sides. Her old LMS number, 7977, has been replaced by her new BR number. Thirty-three of these engines were built for the LMS between 1927 and 1930 for heavy freight duties. This locomotive, built in September 1930, was scrapped in 1956 after working a little over twenty-five years.

Another Beyer-Garratt locomotive. This time, number 47983 is pictured discharging black smoke as she steadily makes her way through Ackworth cutting, West Yorkshire with a long train of empty coal wagons in the summer of 1955.

LMS (LNWR) 0-6-0 Saddle Tank locomotive (Wolverton Works departmental locomotive), number 8, *Earlestown,* pictured at Wolverton in 1956. Despite her apparent good condition, she was withdrawn from service in the following year for scrap.

A total of 260 of these shunting engines were built for the LNWR between 1870 and 1880. They were classified as 'Special Tank Locomotives'. This particular engine spent most of her LMS life working as a departmental shunting engine at Earlestown Locomotive Works in Lancashire, hence her name *Earlestown*. She was later transferred to Wolverton as a Carriage and Wagon Departmental shunting engine, retaining her name. Just five of these engines survived into the BR era, and all were departmental engines at Wolverton. They were all scrapped in the 1950s and none was preserved.

A grim picture of a lone LNER (NER) 'Class J73' tank locomotive standing in the middle of a damp and deserted locomotive shed at West Hartlepool in 1948.

She still displays her old LNER number, 8359, but the letters LNER have been shortened to 'NE' as a wartime economy measure. On the bright side, she did receive her new BR number 68359 and black livery not long after this picture was taken and continued to work at West Hartlepool for over a decade before being scrapped in 1959.

LNER 'Class J63', 'Dock Tank' engine number 68204 pictured in steam outside Immingham Shed in the early 1950s, together with two unidentified LNER 'Class J50' locomotives and a WD locomotive, number 90024.

Built in 1906, number 68204 was the first of just seven dock shunting locomotives designed and built by John Robinson for the Great Central Railway. She spent her entire life working at Immingham Docks before being withdrawn for scrap in 1956.

The Wisbech and Upwell Tramway is the scene of this picture showing an unusual locomotive in the form of an LNER 'Class J70' 'Tram Engine', number 68222. The 'Tram Engine' stands opposite what was its diesel counterpart, number 11102. Both locomotives are engaged in shunting duties. This steam tram engine was one of twelve built for the GER.

The tramway itself was a standard gauge light railway stretching almost 8 miles (12.8 km) from Wisbech to Upwell. It was built by the GER in 1883 for transporting agricultural produce and coal as well as passengers. The railway had eleven large sidings and seven railway stations. The line closed to passenger traffic in 1927 and to freight in 1966.

This superb action photograph shows LMS 'Black Five' locomotive number 45211 piloting 'Jubilee Class' number 45558 *Manitoba* heading a Hull to Liverpool Express passenger train service along the former LNWR 'Spen Valley New Line' near the village of Gomersal, West Yorkshire on 10 October 1959.

The line, which ran between Mirfield and Leeds, closed to intermediate passenger traffic in 1953 and was closed to through express passenger and freight services in 1960, not long after this photograph was taken.

LMS 'Patriot Class' locomotive number 45517 pulls into Castleford Station, West Yorkshire with a York to Manchester Victoria express passenger train on 4 September 1958.

A pair of magnificent LNER 'Class A1' locomotives get ready to depart York Station with their respective express passenger trains during the mid-1950s.

Standing in the foreground is number 60127 *Wilson Worsdell*, whilst number 60138 *Boswell* is standing on the adjacent platform.

LNER 'Class B1' locomotive number 61081 effortlessly ascends Cowlairs incline, Glasgow with a passenger express on 21 June 1949. She is displaying BR numbers and insignia yet is painted in LNER Apple Green livery instead of the official British Railways black livery.

After nationalisation in January 1948, some locomotives were occasionally turned out from locomotive works in England and Scotland painted in pre-1948 livery for a very short period of time in order to use up the stocks of paint which were still on hand in various paint-shops. This was contrary to BR instructions but management usually turned a blind eye. Some of these engines were re-painted in BR livery not long afterwards.

LNER 'Class D16/3' 4-4-0 locomotive number 62562 pictured at Peterborough on 23 July 1957, just three months before she was withdrawn from service for scrap. She was originally built in March 1908 as a GER express passenger locomotive (2P, then re-classified 3P in 1953) but during her later years, she was used for working goods trains.

LMS 'Royal Scot Class' locomotive number 46122 *Royal Ulster Rifleman* pictured at Diggle Station near Oldham whilst working a 9.35am South Shields to Manchester Exchange passenger express in June 1958. Diggle Station closed in 1961.

LMS 'Rebuilt Patriot Class' locomotive number 45521 *Rhyl* passing through Greenfield Station near Oldham with a Trans-Pennine express passenger train in the 1950s.

Fifty-two 'Patriot Class' locomotives were built between 1930 and 1934. Eighteen of them (including number 45521) were later rebuilt into more powerful and efficient engines and were classified as 'Rebuilt Patriot Class' engines. Number 45521 *Rhyl* was first built in 1933 and was rebuilt in 1946.

The dirty black smoke from the chimney and the snowy-white steam from the safety valve adds sparkle to this photograph of an LNER 'Class B16/3' locomotive, number 61454, standing at Scarborough Londesborough Road Station, waiting to depart with the 10am Saturdays-only express passenger train to Manchester in the 1950s.

Londesborough Road Station was an excursion station built by the NER in the 1890s to cope with the large numbers of summer holiday passengers visiting the town. It closed in 1963.

This three-coach passenger train makes a noticeable presence as she hurries through the Essex countryside, churning out an abundance of dirty dark smoke. The culprit, an LNER 'Class B17/6' locomotive, number 61666 *Nottingham Forest*, is working a local stopping service in 1958.

A well turned out 'Class B17/4' locomotive, number 61653 *Huddersfield Town*, pictured at Cambridge Station on 3 June 1951. She is standing quietly at the station platform awaiting passengers to board her local service. In May 1954 this engine was rebuilt as a 'Class B17/6' locomotive and continued in service until 1960.

LNER 'Class B2' number 61671 *Royal Sovereign* hurries through Trumpington with a Cambridge to King's Cross passenger train on 23 April 1951. She is impressively turned out, sporting her fully lined passenger livery.

Built in 1937 as a 'Class B17' locomotive, number 61671 was given the name *Manchester City*. In 1946 she was selected for special duties to work the Royal Train between Liverpool Street and King's Lynn, to convey the Monarch and other members of the royal family to Sandringham House. As a result, she was re-named *Royal Sovereign,* before being rebuilt into a 'Class B2' locomotive in 1948. She then continued to work the Royal Train to East Anglia for over a decade before being withdrawn from service in September 1958 for scrap.

A brand new 2-6-4T 'Class 4P' Fairburn tank locomotive, pictured standing outside Derby Works on 11 May 1946 painted in her new shiny LMS black livery and displaying her LMS number 2226. Having recently left the paint shop, she is waiting to go into service. She was allocated BR number 42226 in 1948 and assigned to Polmadie Shed in Glasgow.

This Fairburn 'Class 4P' locomotive, number 42285, is in immaculate condition as she scurries eastbound through Mirfield, West Yorkshire on 10 August 1961 with some empty coaching stock.

This Ivatt 'Class 2MT' 2-6-2 tank engine, number 41253, of LMS design, was built for BR in 1949. In this picture, she is at Pontefract East Junction, working a local Knottingley to Wakefield Kirkgate stopping train on 12 January 1957.

The line going off to the right of the photograph was a short branch line leading to Pontefract Baghill Station on the York to Sheffield main line. The branch line closed in 1964.

A shabby 0-8-2 tank engine, LMS number 7892, sits neglected in the sidings at Crewe Works. This Bowen-Cooke freight locomotive, built in 1917, was still officially in service when this photograph was taken in 1947 and she was allocated a BR number, 47892. She never wore her new number because she was withdrawn from service in February 1948 and scrapped. Thirty of these engines (Class 6F) were originally built for the LNWR. They were all scrapped.

The tranquil setting of Chorleywood in Hertfordshire is brought to life as this LNER 'Class A3' locomotive, number 60107 *Royal Lancer,* races through the village with the 3.30pm Marylebone to Manchester passenger express on 22 September 1957.

This superb subject photograph, taken at Nottingham Victoria Station in 1950, shows LNER 'Class A3' locomotive number 60102 *Sir Frederick Banbury* in immaculate condition and painted in BR blue livery. This blue livery was only used for a very short period of time from May 1949 until August 1951, when it was tested on the largest of the BR main line passenger locomotives before being discontinued in favour of Brunswick Green.

Nottingham Victoria Station was a former Great Central Railway and Great Northern Railway joint station which opened in 1900. It closed in September 1967.

This close-up shot of another well turned out 'Class A3' 'Pacific' locomotive, number 60063 *Isinglass*, shows her standing at Wakefield Westgate Station whilst working a local Leeds to Doncaster stopping train in the mid-1950s. Her shed plate, clearly visible on the smokebox door, displays the number 35B, indicating that she was working out of Grantham Shed.

Emitting an abundance of black smoke whilst heading north past Pontefract Junction Signal Box, this old Fowler LMS 'Class 4F' locomotive, number 44128, ploughs her way through the Yorkshire countryside with a ten-coach Sunday seaside excursion train to one of the east coast holiday resorts of Scarborough, Bridlington or Whitby on 17 May 1959. Thousands of passengers from the Midlands and the North travelled by excursion train to these and other resorts during summer months.

Built in 1925 for working medium size goods trains, it was not uncommon for these locomotives to be used on passenger trains in later years. They were given the nickname 'Duck Sixes' due to their 0-6-0 wheel arrangements.

Black smoke seems to be the order of the day as another dirty 'Duck Six' locomotive, number 44082, travels north through Ackworth, West Yorkshire with a Whit Monday Bank Holiday excursion train from Sheffield to Scarborough on 6 June 1960.

Black smoke emissions from locomotives were fairly common in the 1950s and 1960s and were usually caused by burning poor quality (usually cheap) coal. Whilst a modern-day environmentalist would be horrified to see it, the spectacle often led to some amazing photographs. Express passenger locomotives at the time usually burned good quality coal which lessened the dirty black smoke effect.

An LMS (LYR) 'Class 2P' 2-4-2 passenger tank locomotive in LMS livery but displaying BR number 50689, is pictured at Pontefract Monkhill Station in 1951 whilst working a local Knottingley to Leeds passenger train. This little tank engine was built back in 1893 for the LYR as a 'Class 5' tank engine and the design was later improved, leading to the development of the more powerful LYR 'Class 6' tank engine which is shown in the next picture.

Another former Lancashire and Yorkshire Railway, 2-4-2 tank locomotive is this LMS 'Class 3P' 2-4-2 passenger tank locomotive number 50909. She is pictured at Normanton Station, West Yorkshire in February 1951. Although displaying her BR number on the side of her coal bunker, she still bears what remains of her old LMS livery.

Built in May 1911 as an LYR 'Class 6' locomotive, she did not quite reach forty years' service as she was withdrawn for scrap a few weeks after this photograph was taken.

A majestic sight to see a newly painted steam locomotive. This LMS 'Class 4P' 3-cylinder compound was photographed at Derby circa 1950 having received her new British Railways black lined livery. She was also proudly displaying her new BR number, 41151.

A total of 195 of these splendid passenger locomotives were built for the LMS between 1924 and 1932. They were almost identical to the earlier Midland Railway 'Class 1000' locomotives but with slightly smaller driving wheels.

This tremendous action photograph shows an LMS 'Class 4P' compound, number 41154, passing through Buxworth in Derbyshire with the 8.27am Manchester Central to Sheffield passenger train on 25 October 1953. Manchester Central Station closed in 1969.

A nice portrayal of an LNER 'Class A8' 4-6-2 Raven/Gresley tank engine, number 69885, pictured at Scarborough in the 1950s. These locomotives were used to work local rural and coastal passenger services until the first generation diesel multiple unit railcars began to replace them in the mid-1950s.

The introduction of diesel railcars onto the network led to the quick demise of these locomotives and all forty-five class members were withdrawn from service and scrapped between 1957 and 1960.

The portrayal of another interesting LNER tank engine is this 'Class T1', number 69918, pictured outside Goole Engine Shed in East Yorkshire, not long before being withdrawn from service for scrap in 1958.

Just ten of these 4-8-0 locomotives were originally built in 1909 and 1910 as powerful (7F), NER 'Class X', shunting engines, designed for moving heavy coal wagons during the loading of coal onto ships at Goole and other similar ports. Goole Docks in the North East of England was built for exporting of coal. A further five of these locomotives (including number 69918) were built by the LNER in 1925 as LNER 'Class T1' locomotives, taking the class total to fifteen.

The words 'dirt' and 'pollution' spring to mind when studying this picture as filthy smoke and steam pour out of this LNER 'Class Q1/2' locomotive as the carefree driver, hand on hip, gazes out of his cab.

The engine is shabby and filthy, with the lettering 'BRITISH RAILWAYS' and the number 69933 almost obliterated by dirt and grime. The photograph was taken circa 1949/1950, the location is Gascoigne Wood Marshalling Yard near Selby and the engine is engaged in shunting duties. Just thirteen 'Class Q1' engines were built and they were all scrapped between 1954 and 1959.

This LNER 'Class V2' mixed traffic locomotive, number 60878, gives off more than her fair share of black smoke as she accelerates out of Pontefract Baghill Station on time, with a busy Newcastle to Cardiff express on 22 December 1959.

LNER 'Class A4' locomotive number 60016 *Silver King* looks a lone figure as she waits patiently on the turntable at Haymarket Shed, Edinburgh in the early 1950s before picking up her coaches then racing up the East Coast Main Line towards London.

Another LNER 'Pacific'. This time 'Class A3' number 60071 *Tranquil* is standing outside Gateshead Shed taking a short break between duties on 30 September 1951. This early photograph was taken when she was still fitted with a single chimney.

LMS & LNER STEAM LOCOMOTIVES: THE POST WAR ERA • 41

This dirty 'Black Five', number 44757, is travelling between Sheffield and York with a Nottingham to Newcastle relief passenger train in 1959.

The unusual design of the front section of the locomotive results from her being fitted with experimental Caprotti valve gear. She was one of twenty 'Class 5' engines to be fitted with the valve gear in 1948. She was also one of just three to receive a double chimney.

This side-on view of LMS 'Class 5' ('Black Five') number 45225 is taken outside York Shed on 3 June 1948 after she stopped to take on fresh water. She was spick and span after having received her new BR black livery.

The new BR 'Lion on Wheel' logo which was displayed on the sides of locomotive tenders did not begin to appear until about 1950, when it replaced the words 'BRITISH RAILWAYS'.

This LNER 'Class D10' locomotive was built in 1913 as a Great Central Railway 'Class E' engine. Just ten were built and were called 'Director Class' locomotives, as each was named after a GCR company director. They were designed as express passenger locomotives.

In this photograph, number 62653 *Sir Edward Fraser* can be seen working a six-coach local passenger train through Knutsford, Cheshire in 1953. All ten class members were withdrawn from service for scrap between 1953 and 1955. This engine was the last surviving class member when she was withdrawn in October 1955.

Another photograph of an LNER 'Class D10' 'Director Class' locomotive. This close-up shot shows number 62654 *Walter Burgh Gair* at Trafford Park Locomotive Shed, Manchester where she was based. The picture is believed to have been taken in the spring of 1950.

All aboard '*The Waverley*' Express! LMS 'Jubilee Class' locomotive, number 45568 *Western Australia*, pictured at Leeds City (Wellington) Station whilst working '*The Waverley*' express passenger service in the late 1950s.

This service consisted of one train per day (Monday to Saturday), in both directions, between London St Pancras and Edinburgh Waverley via Sheffield, Leeds and Carlisle. The down train left St Pancras at 9.10am and the up train left Edinburgh at 10.05am. '*The Waverley*' Express train service ran continually from 1927 until 1969.

A lovely action photograph of another 'Jubilee Class' locomotive, as number 45572 *Eire* speeds into Ackworth cutting, West Yorkshire with the 12.43pm York to Bristol express passenger train service on 16 February 1957.

Number 45572 *Eire* was a regular visitor to this line, as she was one of several 'Jubilee Class' locomotives based at Bristol Barrow Road Shed (22A), which worked daily express services between Bristol, York and Newcastle throughout the 1950s and early 1960s.

'Coronation Class' (Princess Coronation Class) locomotives were the fastest and most powerful locomotives ever built for the LMS Railway Company. This photograph, taken outside Crewe North Shed on 28 August 1949, shows number 46229 *Duchess of Hamilton* painted in LMS 1946 black livery but displaying a BR number that she received in April 1948. She was painted in BR blue livery in April 1950 which was changed to Brunswick Green in April 1952. In September 1958 she was painted in maroon livery.

Number 46229 spent her entire working life as an express passenger locomotive until she was withdrawn from service in February 1964. She was later preserved and is now an important part of the national collection of locomotives owned by the National Railway Museum in York, where she is on public display.

LMS 'Coronation Class' locomotive, number 46242 *City of Glasgow*, works effortlessly non-stop through Lancaster Station whilst hauling '*The Caledonian*' express passenger service to London in 1958. An unidentified 'Black Five' locomotive is creeping into the photograph from the left.

'*The Caledonian*' passenger service departed Glasgow Central in the morning for London Euston. It ran non-stop between Carlisle and London. The train then returned to Glasgow in the afternoon, arriving in the evening. The service operated for just seven years from June 1957 until September 1964.

Pit-props are a thing of the past, used in their thousands when industrial Britain was fuelled by coal. This old former Lancashire and Yorkshire Railway goods engine, number 52411, pictured in 1955, marks her presence with smoke as she saunters through Knottingley towards Wakefield with a consignment of wooden pit-props for use in local collieries.

A total of 484 of these 0-6-0 'Class 27' standard goods engines were introduced onto the LYR between 1889 and 1918. They were absorbed into the LMS and re-classified as LMS 'Class 3F' locomotives. Number 52411 was built in April 1900 and scrapped in 1961.

Another LMS 'Class 3F' former LYR 'Class 27', in need of a good clean, is number 52355, as she stands quietly in some goods sidings at Pontefract on 21 October 1959. She was built in 1895 and toiled endlessly for over sixty years before being scrapped in 1961.

LMS 'Class 4P' Fowler 3-cylinder compound number 40931 pictured at Hellifield Station in North Yorkshire working a local passenger train service to Morecambe in the early 1950s.

Another Fowler 'Class 4P' locomotive, number 40937, pictured as a light engine outside Bank Hall Shed near Liverpool in the early 1950s. She was assigned to Bank Hall Shed at the time but was later transferred to Lancaster Green Ayre Shed before being withdrawn from service in April 1958 for scrap.

LNER 'Class Q6' locomotive number 63395 crawling through Pontefract, tender first, on 9 March 1957 destined for either Frickley or Goldthorpe Colliery with some empty coal wagons.

She was built in 1918 for the NER Company as a 'Class T2' heavy freight locomotive (power classification 6F) and after being withdrawn from service in 1967, was preserved and can be seen working on the North Yorkshire Moors Heritage Railway.

This LNER 'Class J39/1' mixed traffic, medium powered (4P5F) locomotive, number 64760, is assisted by a banking engine at the rear of her heavy coal train as she ploughs her way through Scunthorpe in 1950. A total of 289 of these Gresley designed engines were built between 1926 and 1941 and worked all over the former LNER network from London to Scotland.

Another LNER 'Class J39/1' locomotive is number 64721, pictured having just left the Vale of York with an up 'Class D' express freight train in 1955.

'Class J39' locomotives were sub-divided into three sub-classes: 'J39/1', 'J39/2' and 'J39/3'. The only differences related to the tenders which the engines were fitted with. This photograph and the previous one both show 'Class J39/1' locomotives fitted with LNER standard 3,500 gallon tenders. 'J39/2' locomotives had LNER standard 4,200 gallon tenders and 'J39/3' engines were fitted with various former NER tenders.

A tranquil setting, showing an LNER 'Class J17' locomotive, number 65557, standing in Melton Constable Station, Norfolk, whilst working a 'Class H' (through freight) goods train on 13 September 1951.

The driver and fireman appear relaxed in the cab of the engine, as the rather portly station shunter (hand in pocket), walks towards them with his shunting pole. The engine was based at Melton Constable.

An LMS 'Crab' locomotive, number 42715, is pictured working a passenger express on the Spen Valley New line, non-stop through a deserted Battyeford Station near Mirfield, West Yorkshire on 12 September 1953. Battyeford Station had closed in January of that year.
The line in question, the 'Spen Valley New Line', opened in 1900 as a duplicate line between Leeds and Huddersfield. All stations along the line were closed to passenger traffic in 1953 and the line itself closed to both freight and through express passenger traffic in 1960.

Another LMS 'Crab', number 42944, is captured on camera whilst performing passenger coach shunting duties at Euston Station, London in August 1955.
LMS 'Crab' locomotives, also known as 'Horwich Moguls', were easy to identify and were noted for their highly angled cylinders. They were excellent mixed traffic locomotives and well-liked by their crews. A total of 245 were built between 1926 and 1932, and number 42944 in this picture was the last member of the class to be built.

A chinwag in the cab of this LMS 'Class 2F' 0-6-0 tank engine, number 58850, as the driver, fireman and shunter are engrossed in conversation whilst taking a short break from shunting duties on the Cromford and High Peak Railway line in the early 1950s.

The engine was originally built as a docks shunting engine for the NLR in 1880. Later loaned to the LNWR before being absorbed into the LMS in 1923, where she worked as a shunting engine on the High Peak Line in Derbyshire until 1960, when she was withdrawn from service and later preserved. She is now a part of the Bluebell Heritage Railway in West Sussex.

Built in 1888, this 0-6-2 'Webb Coal Tank', number 58926, takes on water outside Merthyr Tydfil LMS Shed on 6 January 1958, having worked behind pilot engine number 49121 the previous day on a double-headed special passenger train organised by the Stephenson Locomotive Society. The train was the last passenger train to run on the former Merthyr, Tredegar and Abergavenny Railway line, after its closure to passenger traffic the day before. The special train was to commemorate the event. The line was later closed completely.

Although this was a sad occasion for railway enthusiasts, the future of engine number 58926 was more assured. After being withdrawn for scrap in 1958, she was purchased privately for preservation and is now working on the Keighley and Worth Valley Heritage Railway

LMS un-rebuilt (at the time) 'Royal Scot Class' locomotive number 46137 *The Prince of Wales's Volunteers South Lancashire,* newly painted in her early BR black livery, whilst standing outside Crewe North Shed on 8 May 1948. She was originally named *Vesta* after being built in 1927 but was renamed in 1936.

All seventy class members were rebuilt between 1943 and 1955, with number 46137 in this picture being the last and only one to receive her rebuild in 1955. She spent almost all of her BR service working express passenger train services whilst based at Carlisle Upperby Shed.

A 'Royal Scot' in action. This time, number 46142 *The York and Lancaster Regiment* hauls an express passenger train through Scout Green in Cumbria on the West Coast Main Line. Scout Green Signal Box is on the right of the picture, which was taken in 1959.

Scout Green was a popular location for photographing steam locomotives as they often produced spectacular pictures with plenty of smoke as locomotives worked hard ascending the 1 in 75 gradient towards Shap.

This superb photograph shows LNER 'Class A4' number 60031 *Golden Plover* in pristine condition, having just left the paint shop at Doncaster Works on 3 June 1948. She is turned out in BR white lined 'Express Blue' livery. During this period, BR was conducting 'livery trials' to ascertain the most suitable liveries to be used on British Railways locomotives.

A different 'Class A4' livery is depicted in this picture. The famous 'Garter Blue' livery was introduced in May 1937 for use on 'Class A4' locomotives and continued to be used after the war until 1948. During the war years, all locomotives were painted black.

This photograph shows number 60001 *Sir Ronald Matthews* passing St Margaret's, Edinburgh with an express passenger train on 24 June 1949. She is painted in garter blue livery. Originally named *Garganey* when built, she was renamed *Sir Ronald Matthews* in January 1949, not long before this picture was taken.

Another LNER 'Class A4' locomotive? Close but no cigar. Take another look and you will see that this locomotive does not have the 'Class A4' 4-6-2 wheel arrangement but has a 4-6-0 configuration. She is in fact an LNER 'Class B17/5', number 61659 *East Anglian*. This photograph was taken outside Cambridge Shed in 1949.

Just two 'Class B17/4' locomotives were rebuilt into 'Class B17/5' engines in 1937 and given a streamlined casing similar to but shorter than that of the 'Class A4' engines. The streamlined appearance of these locomotives was short lived as the casing was removed from both engines in April 1951, making this photograph quite rare.

When built as a 'Class B17' in 1936, number 61659 was named *Norwich City*, but was re-named *East Anglian* in 1937 when she was streamlined to work the prestigious '*East Anglian*' express passenger train service which operated between London (Liverpool Street) and Norwich.

A nice clean Ivatt 'Class 2' locomotive, number 46413, passing Prince of Wales Colliery, Pontefract with a local two-coach Leeds to Knottingley passenger service in October 1957.

These small locomotives were built as light goods engines for the LMS but were later re-classified by BR as 'mixed traffic' locomotives (2MT) and were frequently used for working light passenger and local branch line services. Due to their small size they earned the nickname 'Mickey Mouse'.

Another 'Mickey Mouse' Ivatt 'Class 2' engine approaches the Prince of Wales Colliery from the opposite direction to the last picture. This time, number 46453 is working a Knottingley to Leeds local passenger stopping train on 15 March 1958.

LNER 'Class J6' light goods engine number 64205, pictured near Runtlings Junction, Ossett near Wakefield, working an up goods train during the 1950s. Over 100 of these locomotives were originally built at Doncaster in 1911 for the GNR.

An unidentified LMS 'Class 8F' locomotive pictured with two guard's vans at Wrangbrook Junction near Upton, West Yorkshire in 1957. Wrangbrook Junction is on the former Hull and Barnsley Railway line where two separate branch lines divert from the H&B line.

The locomotive pictured in the photograph is travelling on the Denaby branch line whilst the empty wagons in the top of the picture are stabled in some sidings on the Wath branch line.

LNER 'Class O4/7' locomotive number 63570, pictured outside Tuxford Shed, Nottinghamshire on 1 August 1949. Although her new BRITISH RAILWAYS lettering is painted on the tender and her BR number is displayed on the cab-side, her condition leaves a lot to be desired and a good lick of paint would not go amiss.

Despite her appearance in this picture, she continued working for over a decade before she was withdrawn for scrap in December 1961.

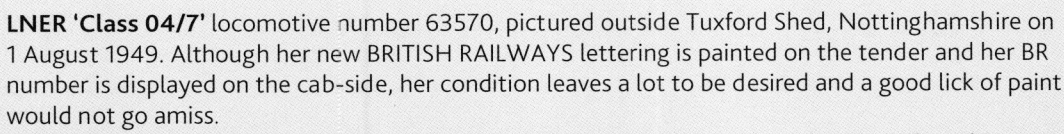

A nice plain side view of an LNER 'Class Q5/1' locomotive standing in some sidings outside Borough Gardens Shed at Gateshead on 25 June 1950. Built at Darlington in 1911 for the NER to work heavy mineral traffic (6F) she was on her last legs when this picture was taken.

Although still displaying her old LNER number, 3326, she had been allocated her BR number 63326 but never wore it as she was withdrawn from service for scrap in October 1951.

A Scottish setting is the theme for this picture as LNER 'Class K4' locomotive number 61995, *Cameron of Lochiel,* arrives at the beautiful town of Fort William in the Western Scottish Highlands with a special passenger working on the 18 June 1960.

These 'K4' locomotives, were specifically designed by Nigel Gresley to work trains on the steep gradients and tight curves of the former NBR West Highland line from Glasgow to Fort William and Mallaig.

The Scottish theme continues with two locomotives, double-heading the 10.24am, Fort William to Mallaig passenger service as it enters Glenfinnan Station on 11 June 1959. The pilot locomotive is an un-named LNER 'Class K1', number 62012, which is piloting an LNER 'Class K1/1' engine, number 61997 *MacCalin Mor*.

Number 61997 was classified a 'Class K1/1' engine because she was originally a 'Class K4' 3-cylinder engine rebuilt in 1945 as a 2-cylinder 'Class K1' prototype for the new Peppercorn 'K1' Class. Number 61997 was classified as the sole 'Class K1/1' locomotive and the seventy new Peppercorn locomotives were then classified as LNER 'Class K1' engines.

Another 'Class K1' visiting the Scottish Highlands. In this picture, Fort William based locomotive number 62052 is seen piloting a 'Class B1', number 61243 *Sir Harold Mitchell*, on a passenger express at Fort William Station in June 1960.

Number 62052 was the first 'Class K1' to be withdrawn from service. She was withdrawn on 28 February 1963 and later scrapped at Cowlairs Works in Glasgow.

Having viewed several LNER engines in Scotland, time to switch to some of the LMS locomotives making their presence north of the border. This rather unusual yet interesting photograph shows 'Coronation Class' locomotive number 46222 *Queen Mary* having arrived at her destination, Glasgow Central, on 28 October 1955.

Glasgow Central was the terminus station for these 'Coronation Class' and other express passenger locomotives working the West Coast Main Line passenger trains from London Euston to Glasgow.

LMS 'Class 3F' number 57626, an old former Caledonian Railway goods engine (CR No.291), is pictured at Beattock, meandering her way through the sparse countryside of Dumfries and Galloway, on 04 August 1956.

This rare photograph shows a former Scottish Highland Railway 'Clan Class' locomotive (LMS 'Class 4P'), number 54767 *Clan Mackinnon* (former HR number 55), at her home shed of Inverness, immediately prior to being withdrawn from service in January 1950.

Just eight of these 4-6-0 locomotives were built, and withdrawals started in 1943. Number 54767 was the sole remaining member of the class when this photograph was taken. They were all scrapped.

LMS 'Royal Scot Class' locomotive number 46113 *Cameronian*, pictured at Dumfries Station in Scotland, working a down Glasgow express passenger service in 1959.

LMS 'Coronation Class' locomotive number 46220 *Coronation*, pictured at Glasgow Central Station on 28 October 1955, waiting to depart for London with the famous '*Royal Scot*' express passenger train service.

A shed plate affixed to the smokebox door of the locomotive is displaying the number 66A, which shows that the engine is based locally at Glasgow Polmadie Shed.

A 'Black Five', number 44722, passes the signal box as she pulls into The Mound Junction Station in Scotland with the 10.40am Inverness to Wick passenger service on 17 June 1959.

The Mound Junction Station, near the head of Loch Fleet, was originally part of the Sutherland Railway Company, which was absorbed into the Highland Railway in 1884. The station closed in 1960.

The LMS Railway did not just cover England and Scotland as this Stanier 'Class 5' ('Black Five') locomotive, number 44738, shows when photographed near Llandudno Junction in North Wales whilst working an express passenger train in the 1950s.

The unusual triangular shaped front of the engine was due to experimental Caprotti valve gear, which was fitted to just twenty 'Class 5' engines (numbers 44738–44757). The valve gear proved to be just moderately successful. Number 44738 was built at Crewe in June 1948 and started work as a BR locomotive assigned to Llandudno Junction Shed, close to where this picture was taken.

This side view photograph shows another 'Class 5' ('Black Five') locomotive, number 44755, fitted with the experimental Caprotti valve gear. In addition, she was one of just three (44755–57) to be additionally fitted with a double chimney, which is clearly visible in this picture.

The photograph was taken outside Crewe Works on 8 May 1948, not long after she emerged from the paint shop in black BR livery as a brand new locomotive destined for Derby Shed, where she had been assigned to work.

Coal was without doubt the biggest single freight commodity transported on Britain's Railways during the days of steam, and large amounts of it were used to fuel the railway industry. Millions of coal wagons, both full and empty, were transported all over the country every year. Railway companies such as the Taff Vale Railway in South Wales made vast fortunes transporting coal, which they dubbed 'black gold' or 'black diamonds'.

This picture is a typical example of a coal train operating in the 1950s. LMS 'Class 8F' locomotive number 48641 is seen here having departed the Prince of Wales Colliery in Pontefract, West Yorkshire with a local coal train heading a short distance to her destination at Hunslet, Leeds. This photograph taken in 1955 also shows part of the POW Colliery in the background.

A high vantage point makes for an interesting photograph of this 'Class 8F', number 48201, as she ascends Gunness Bank in Lincolnshire, discharging grey/black smoke as she makes her way eastbound towards Scunthorpe with a heavy coal train in the 1950s.

Whitworth Colliery near Normanton in West Yorkshire is the scene of this picture, as a 2-8-0, Stanier 'Class 8F', locomotive number 48622 departs the colliery, tender first, with a local coal train on 4 May 1957.

LNER 0-6-2T 'Class N7/3' locomotive number 69692, pictured at Annesley, Nottinghamshire, working a workman's train known as the '*Annesley Dido*' in the 1950s. A railway employee can be seen alighting from the train at an isolated line side location, presumably near his home.

'Dido' trains were special 'staff trains' which transported railway staff to and from work on a daily basis, 'day in day out', hence the name dido. They were used for staff working at locations (often remote locations) at times when other forms of public transport were unavailable. Motor cars were seldom owned by working class families at the time.

The '*Annesley Dido*' in this photograph ran from the 1920s until 1962, primarily for staff working at Annesley engine shed, approximately twelve miles outside Nottingham.

Another LNER tank engine is this 'Class F6' 2-4-2T locomotive, number 67238, pictured passing Thetford Signal Box, having just left Thetford Station with a local passenger train service to Bury St. Edmunds on 3 June 1952.

The railway line from Thetford in Norfolk to Bury St. Edmunds in Suffolk was originally built as the Bury St. Edmunds and Thetford Railway in 1876 and later sold to the GER. The line closed to passenger traffic in June 1953, just a year after this picture was taken but continued to be used for goods traffic until it closed completely in 1960.

This very old-fashioned looking tank engine is an LNER 'Class C12' 4-4-2T locomotive, still displaying her LNER number 7356 in May 1948 when this picture was taken. She is leaving Beeston Station in West Yorkshire with a local Castleford to Leeds passenger train. The letters 'LNER' from the side of her tank have been shortened to 'NE'. This practice was carried out on LNER engines after 1941 as a wartime economy measure.

LNER 7356 (BR 67356) was built at Doncaster for the Great Northern Railway in 1900, and by sheer coincidence an old GNR 'Home' Somersault Signal appears on the extreme right-hand side of this picture. In 1948, number 7356 was spending her final years based at Leeds, Copley Hill Shed, before being withdrawn from service and scrapped in October 1951.

From Leeds, we move north of the border back to Scotland for a fine close-up of this LNER 'Class C15' 4-4-2 tank engine, number 67464, standing in Grangemouth Station near Falkirk in Stirlingshire waiting for the guard to wave his green flag to signal the departure of this local passenger service.

The photograph was taken in 1949 before the name 'BRITISH RAILWAYS' was replaced by the 'Lion on Wheel' emblem on the side of the cab. The name 'Polmont', painted on the front buffer beam, is the name of a locomotive shed near Falkirk where the engine was based.

Aberdeen Docks is a splendid backdrop to observe the crews of these two LNER shunting engines having a little natter during a short fag break on 22 September 1955.

The locomotives on view are two 0-4-4T, shunting engines. Number 68190 (left) is an LNER 'Class Z4' and number 68192 is an LNER 'Class Z5'. Both engines were originally built for the Great North of Scotland Railway Company, specifically to perform shunting duties at Aberdeen Docks. This photograph must be considered quite rare, as only two members of each class were ever built. All four engines were built in 1915 and worked at Aberdeen Docks until they were withdrawn from service for scrap between 1956 and 1960.

LNER (NER) 'Class B16/2' locomotive number 61421, powers her way through Pontefract with an express goods train, heading towards Sheffield on 25 October 1958.

Seventy of these highly successful 'B16' locomotives designed by Vincent Raven were originally built at Darlington between 1919 and 1924 as NER 'Class S3' mixed traffic engines.

A hot summer's day in the mid-1950s sets the scene for this picture, as the driver of a double-header seaside express waits patiently in York Station for his cue to depart. The train, jam-packed with holidaymakers, is bound for the popular north east coastal holiday resort of Scarborough. The pilot locomotive heading the train is an LNER 'Class B16/1', number 61441.

This LNER 'Class B17/1', number 61630 *Tottenham Hotspur,* pictured at Doncaster, is in immaculate condition and ready to be put back into service after having received an overhaul at Doncaster Works just prior to this photograph being taken on 19 October 1956.

Sadly, this would be her last major overhaul as she was withdrawn from service and scrapped in August 1958, less than two years after this picture was taken.

This nice portrayal of LNER 'Class B1' locomotive number 61319 was taken on 5 June 1958, outside Borough Gardens Shed, Gateshead, where she was based. Although displaying her new 'BRITISH RAILWAYS' lettering on the tender and sporting her BR number on the cab-sides, she is not painted in standard BR black livery. Instead she is painted with the much more splendid LNER apple green livery, which is a sight to behold.

A 'Class B1' in action, as number 61016 *Inyala* passes over the level crossing at Market Weighton Station in East Yorkshire with a passenger train in May 1952. These mixed traffic locomotives were extremely efficient and ideal for working both passenger and freight services.

Producing an abundance of grey/black smoke is this 'Class B1' locomotive, number 61033 *Dibatag*, as she heads out of Pontefract on her southbound journey with a York to Bournemouth express passenger train on 7 June 1958.

A nice interesting photograph of an Ivatt design 2-6-2T 'Class 2P' tank engine (BR 2MT), number 41204. Her shed plate, displaying the number '84H', is clearly visible on the smokebox door, showing that she was based at Crewe Gresty Lane Shed where this photograph was taken in 1960.

Gresty Lane Shed closed in 1963 and number 41204 was transferred to Stockport until she was withdrawn from service in 1966. She was later scrapped.

LMS Ivatt 'Class 2P' (2MT) locomotive number 41253 is working the 12.25pm Goole to Wakefield Kirkgate local passenger service on 13 June 1957.

The train, consisting of just three coaches, is pictured leaving Knottingley Station which is visible in the background.

Another photograph of 'Class 2P' (2MT) locomotive number 41253 working a local stopping train from Knottingley to Wakefield. On this occasion she is working a four-coach, Saturdays-only service through Pontefract Monkhill on 23 March 1957.

Pushing the Pontefract turntable was no easy task for one man, but necessary to prepare this LNER 'Class Q6' goods engine, number 63406, for her next task of working a local goods train along the Brackenhill Branch line at Ackworth in West Yorkshire on 13 May 1957.

The Brackenhill Branch line was in fact a light railway, built in 1901 from Brackenhill Junction, Ackworth (on the Sheffield to York line) to Hemsworth Colliery, three miles away. The line was never used or intended for passengers but served Hemsworth Colliery and a number of sandstone and limestone quarries in the vicinity. The line was absorbed into the LNER in 1923 and closed in 1962, after which, the track was removed. Hemsworth Colliery closed in 1969.

Hemsworth is the location of this photograph, although the colliery is out of view, as an LNER 'Class O4/7', number 63857, hauls a fully laden coal train on the Leeds to Doncaster main line, towards Doncaster, on 12 May 1959.

The Summer Solstice sets the scene for another coal train as it trundles through Hemsworth towards Doncaster. This time a 'Class O4/8' locomotive, number 63893, provides the power to haul this heavy mineral train on 21 June 1960. Having taken this picture, the photographer relaxes and basks patiently in the glorious summer sunshine waiting to capture his next shot.

A grubby looking LNER 'Class O4/6' number 63906 is pictured at Scunthorpe in the mid-1950s. She was based at Frodingham Shed, on the outskirts of the town.

It is an overcast day at Altofts and Whitwood Station near Normanton in West Yorkshire in 1958, as an elderly couple on an otherwise deserted station stand well back from the platform edge as their train arrives.

The local train they have been waiting for is headed by an LMS Fowler, 4-4-0 'Class 2P' locomotive, number 40581, which pulls slowly into the station.

A total of 138 of these light passenger engines were built for the LMS between 1928 and 1932 and were ideal for this type of light passenger work.

LMS & LNER STEAM LOCOMOTIVES: THE POST WAR ERA • 73

This close-up shows another Fowler 'Class 2P' locomotive, number 40602. This photograph was taken on 13 August 1950 at Carlisle Kingsmoor Shed where the engine was based. She was not in steam at the time although her tender is filled with coal ready for use.

Number 40602 is in early BR livery with the words 'BRITISH RAILWAYS' painted on both sides of the tender, a temporary measure used in 1948 and 1949 before the 'Lion on Wheel' logo was introduced.

This plain and simple photograph is a good example of an LMS (MR) 4-4-0 'Class 2P' locomotive and makes an ideal picture for an enthusiast to study.

The picture features a former Midland Railway locomotive, number 40556, not in steam, standing outside Hasland Engine Shed, Chesterfield, on 8 July 1951. She appears to be in splendid condition, painted in British Railways lined black livery and displaying the first BR 'Lion on Wheel' emblem on her tender. The tender itself is full of what appears to be good quality shiny black coal. Altogether an imposing picture.

LMS 'Patriot Class' locomotive number 45519 *Lady Godiva* pictured as a light engine at Pontefract on Thursday 30 July 1959. These engines were built by the LMS as express passenger locomotives with a power classification of 5XP (between 5P and 6P). They were, however, re-classified by BR in 1951 as mixed traffic locomotives and given a new power classification of 6P5F.

A total of fifty-two 'Patriot Class' engines were built between 1930 and 1934. Eighteen of these were rebuilt between 1946 and 1948 into a new class of locomotive designated as 'Rebuilt Patriot Class'. The remaining thirty-four (including this one) were thereafter generally referred to as 'Un-rebuilt Patriot Class' locomotives.

This picture is of a 'Rebuilt Patriot Class' locomotive, which clearly differs from number 45519 shown in the previous picture (above).

Taken at Crewe in 1948, the photograph shows 'Rebuilt Patriot Class' locomotive number 5530 *Sir Frank Ree* still painted in LMS livery and displaying her LMS number. She had been rebuilt just two years earlier when she was converted from a 'Patriot Class' engine to a 'Rebuilt Patriot Class'.

Sir Frank Ree was a former General Manager of the NLR and the LNWR.

Newcastle Station is the setting for this photograph as LNER 'Class A3' locomotive number 60067 *Ladas* stands alongside the platform, having arrived at her destination with an express passenger train from King's Cross circa 1961/62. Passengers have alighted from the train and one of the head-code lamps has been removed from the front of the engine.

Number 60067 is fitted with the German type 'Witte' smoke deflectors. They were first tested by BR in October 1960 and after the tests were deemed successful, they were fitted to 'Class A3' locomotives from March 1961. The smoke deflectors in this photograph were fitted to number 60067 in July 1961, shortly before this picture was taken. Sadly, this magnificent engine was withdrawn from service for scrap in December 1962.

Another 'Class A3', number 60055 *Woolwinder,* is pictured hurtling along a long straight stretch of the East Coast Main Line, through Hadley Wood near Potters Bar, whilst hauling an express passenger train on 4 July 1953.

Standby duties are the orders of the day for this LNER 'Class A3' locomotive, number 60076 *Galopin*, pictured outside Darlington Shed in the 1950s. She is in steam and her crew are on hand to take her out at a moment's notice.

During BR steam days, standby 'Pacific' locomotives were used at various locations on the East Coast Main Line as 'back-up' engines to take over express train workings in the event of any engine failures. This ensured that passenger train disruption was kept to a minimum.

LNER 'Peppercorn design' 'Class A1' locomotive number 60130 *Kestrel*, pictured at King's Cross Station in London, basking in the sunshine whilst preparing to depart with an express passenger train service to Leeds in the late 1950s.

LNER 'Thompson design' 'Class A2/1' locomotive, number 60508 *Duke of Rothesay,* is pictured racing through Heck Station near Selby in North Yorkshire on the East Coast Main Line with the down '*Heart of Midlothian*' express passenger train on 31 July 1958.

Just four 'Class A2/1' locomotives were built by the LNER at Darlington in 1944, as mixed traffic locomotives (7P6F).

The '*Heart of Midlothian*' express passenger service operated between London King's Cross and Edinburgh Waverley from 1951 until 1968.

Heck Station closed to passenger traffic on 15 September 1958, shortly after this photograph was taken. It closed to goods traffic in April 1963 and was later demolished.

'Class 3F' 0-6-0 'Jinty' tank locomotive LMS number 7406, pictured in her LMS livery whilst performing shunting duties at Carnforth on 26 April 1947. The railways were nationalised on 1 January 1948, after which, number 7406 was allocated her BR number 47406 before being repainted in BR black livery.

Built in December 1926, she spent a short time working at Warrington and Crewe before moving to Carnforth Shed (11A) where this picture was taken. For the next thirty-two years she was at Carnforth, engaged in shunting operations.

After being withdrawn from service in 1966, she spent almost seventeen years at a scrapyard in South Wales before being rescued for preservation. After a long period of restoration she returned to steam in 2010 and is now fully operational on the Great Central Heritage Railway.

Another locomotive painted in LMS livery, albeit in a dilapidated condition, is this 'Class G2A' heavy freight (7F) locomotive, number 8921. She is standing outside the Engine Shed at Liverpool Edge Hill, circa 1947/48.

Number 8921 (later BR 48921) was built at Crewe right back in 1902 for the LNWR. These engines were rebuilds of earlier 'Class G1' locomotives and they were manufactured in huge numbers. In total, 327 were built and 320 of them entered BR stock in 1948. They were given the nicknames 'Super Ds' or 'Duck Eights'. Number 8921 (BR 48921) continued working for BR until April 1958, when she was withdrawn from service for scrap.

This LMS 'Class G1' locomotive, number 49089, seems to be out of action as she stands in the yard outside Crewe Works, circa 1950. She is painted black and displaying a new BR number on her cab side. She appears to be in quite good condition.

Although the exact date of this photograph is not known, records show that number 49089 was given her BR number in January 1950 and assigned to Aston Shed in Birmingham. She was withdrawn from service just three months later in April 1950 and then then 'mothballed' until she was scrapped at Crewe Works in 1954.

LMS Livery graces this old 'Class 3F' locomotive, number 12435, as she chugs along through Heaton Lodge Junction, Mirfield, pulling a local goods train on 12 May 1947.

Built in 1901 as a Lancashire and Yorkshire Railway 'Class 27' locomotive, number 12435 (later BR 52435) is a fine example of such an engine. A total of 484 of these 2-cylinder locomotives, designed by John Aspinall, were built between 1889 and 1918 and were the standard goods engines for the LYR before being absorbed into the LMS.

LMS Ivatt 'Class 2P' (BR 2MT) tank locomotive number 41251, pictured at Pontefract whilst working a Saturday-only local passenger train to Wakefield in 1957. These engines were given the nickname 'Mickey Mouse' due to their small size.

Another 'Mickey Mouse' engine, pictured arriving at Uppingham Station in Rutland with the 2.49pm arrival from Seaton on 8 October 1956. Uppingham was a terminus station on the branch line from Seaton until it closed in 1960.

The locomotive hauling the train is an Ivatt 'Class 2P' (BR 2MT) engine, number 41278. She was one of fifty class members (out of a total of 130) to be fitted with 'Push/Pull' apparatus for working on branch lines such as this one.

LNER 'Class D10' (Director Class) locomotive number 62650 *Prince Henry*, pictured at Manchester Central Station about to depart with the 12.35pm local passenger train service to Northwich, Cheshire, on 4 April 1953. The engine was based at Northwich Shed at the time.

When built in 1913, this former GCR 'Class 11E' express passenger locomotive was given the name *Sir Alexander Henderson,* who at the time was the Chairman of the board of directors for the Great Central Railway, but the name was later changed. She carried the name *Prince Henry* from 1920 until she was withdrawn from service in 1954.

Spouting black smoke as she ploughs her way through a cutting, this 'Large Director Class' 'D11/1' locomotive, number 62666 *Zeebrugge*, hauls a York to Swansea express passenger train through the Yorkshire countryside on 28 June 1957.

This former GCR 'Class 11F' locomotive, designed by John Robinson and built in 1922, was a development of his 'Class 11E' (LNER 'Class D10') locomotive, an example of which appears in the previous photograph.

LNER 'Class D16/3' locomotive number 62543, pictured standing in March Station, Cambridgeshire with a local passenger train on 8 April 1958.

This former GER locomotive, with its stylish, sophisticated design, was built in 1903 and during the height of the Edwardian Era would have been scurrying along the Great Eastern Main Line hauling the latest express passenger trains whilst adorned in GER blue lined livery.

A snowy winter's day in January 1958 sets the scene for this eye-catching photograph taken at Holbeck Low Level in Leeds as two LNER 'Class D49/2' 'Hunt Class' locomotives cross paths on their respective journeys.

In the centre of the picture, number 62756 *The Brocklesby* gathers speed as she heads out of Leeds with an express passenger train bound for Harrogate, whilst number 62740 *The Bedale* enters the picture on the left with her express from Harrogate, reducing speed as she approaches her destination in Leeds.

LNER 'Class D49/2' 'Hunt Class' number 2737 *The York and Ainsty* exposing remnants of her LNER livery, is in a woeful state outside her home shed of Hull, Botanic Gardens in 1947. Although designed primarily to work intermediate express passenger trains, she had been put to use on freight duties.

A few months after this photograph was taken, the railways were nationalised and 2737 was re-painted in new BR livery. With her nameplate gleaming and her new number 62737 displayed, she continued working out of Hull for over a decade before being withdrawn from service for scrap in January 1958.

LMS 'Class 2P' former Midland Railway, Johnson design, 4-4-0 locomotive number 40538, heading south through Pontefract on 19 May 1956 with a four-coach, Saturday-only, York to Sheffield local stopping train.

Built at Derby in 1899, number 40538 was used to work express passenger trains in her early years. However, she was well into her twilight years when this picture was taken, although she continued slogging it out for another three years before being withdrawn from service for scrap.

LMS 'Class 2P' Fowler design 4-4-0 locomotive number 40613 pictured in steam outside Kingsmoor Shed, Carlisle in August 1950, displaying her early BR black livery.

A total of 138 of these light passenger engines were built at Derby and Crewe between 1928 and 1932. They were designed by Henry Fowler and based upon the former Midland Railway '483 Class' locomotives. Number 40613 in this picture survived until 1961.

Another LMS 'Class 2P' stands quietly in Ashchurch Station, Gloucestershire in September 1957 whilst operating a local single coach push and pull service.

The locomotive in this picture is a Stanier design 0-4-4 tank engine, number 41900, which was built in 1932 for light passenger duties. She was fitted with push and pull apparatus in September 1950.

Just ten of these tank engines were built. This was the first member of the class to be built and she survived the longest, working until 1962. The other nine were all withdrawn from service in November 1959.

This intriguing photograph features an Ivatt 'Class 2P' (BR 2MT), 'Mickey Mouse' 2-6-2 tank engine, number 41206, standing on a remote turntable in bleak surroundings at Garsdale, on the Settle to Carlisle line, high in the Yorkshire Dales.

The turntable is surrounded by a palisade of wooden railway sleepers for protection from the strong Pennine winds. This was deemed necessary after a locomotive standing on the turntable was spun around by strong winds in 1900.

LNER 'Class J11/3' locomotive number 64417 pictured at Pontefract on 16 June 1956, working a Manchester London Road to Scarborough, Saturdays-only, summer passenger service.

Built in 1907 as a GCR goods engine, this LNER 'Class J11' was designed for freight work (3F) but it was not uncommon, particularly during summer months, for these locomotives to be used on passenger services such as this one. In 1956, number 64417 was working out of Barnsley Shed and would have pulled this train on the middle leg of her journey from Penistone or Barnsley via Wath Junction to South Milford. The train would have then continued via Selby and Bridlington to Scarborough.

LNER 'Class J37' locomotive number 64639 rattles through the tiny station of Craiglockhart in the suburbs of Edinburgh with a train load of empty wagons on 15 May 1954. The engine is a former NBR 'S Class' freight locomotive. Originally allocated a power classification of '4F' by the NBR and the LNER, she was re-classified as a '5F' engine by BR in the 1950s. Craiglockhart Station was closed in September 1962.

Edinburgh in the 1950s is also the setting for this picture, as an LNER 'Class V1' 2-6-2T Gresley design tank locomotive, number 67629, works her way through Haymarket with a local three-coach passenger train. It is a pleasure to see a locomotive and her coaches in such pristine condition.

Yet another spick and span tank engine. This time, LNER 'Class A5/2' locomotive number 69842 is pictured at Darlington Station, waiting to take a local passenger train to the seaside resort of Saltburn near Redcar in the 1950s. These engines were based upon the design of GCR 'Class 9N' locomotives, which were purpose built for working suburban passenger trains like this one. The engine in this picture was based locally at Darlington Shed (51A).

LMS & LNER STEAM LOCOMOTIVES: THE POST WAR ERA • 87

This tank engine is in full flow, letting off steam and smoke as she scurries along with the 4.18pm local passenger service from Keighley towards Ingrove on 5 June 1954. The train would have been destined for Bradford or Halifax.

The locomotive heading this train is LNER 'Class N1' 0-6-2 tank engine number 69485, built at Doncaster in 1912. She was withdrawn from service in November 1954, not long after this picture was taken. She was later scrapped.

This interesting photograph shows an old GNR brick-built warehouse at Grimsby Docks in the 1950s. In front of the warehouse stand three LNER 'Class J94' 'Saddle Tank' engines in a shabby condition and looking the worse for wear.

The engine on the extreme right is number 68068. In front of her stands number 68071, with her fireman crouching on the top of the engine in the days before Mr Health and Safety appeared on the scene. A third unidentified 'Class J94' engine is sandwiched between the warehouse and a box van. All three engines are engaged in shunting duties.

A total of 344 of these 0-6-0 saddle tanks were built for the Government War Department, for wartime use as shunting engines during the Second World War. At the end of the conflict seventy-five were purchased by the LNER, who classified them as 'Class J94' locomotives.

LMS 4-4-0 'Class 4P' Fowler 3-cylinder compound locomotive number 41152 is pictured in the sidings at Lancaster Green Ayre shed, where she was based when this photograph was taken on 24 March 1957.

Although seemingly in quite good condition, she was entering her final year of service before being withdrawn for scrap on 31 March 1958.

Destination unknown, this Fowler 'Class 4P' compound, number 41105, based at Rugby, whizzes through Watford Junction with a down express passenger train in the early/mid 1950s.

A London Transport, (Bakerloo Line), underground train is just visible in the distance on the extreme right of the picture.

Number 41105 finished her working life in 1956 and was 'mothballed' at Rugby Shed in October of that year. She was formally withdrawn from service on 30 September 1957 and scrapped at Derby Works shortly afterwards.

A nice action photograph of an LMS Fowler 'Class 4P' 2-6-4 tank locomotive, number 42333, letting off steam through her safety valve as she works a local passenger train near Trent Junction, Nottingham in the 1950s.

Nice side view photograph of this LMS Fowler 'Class 7F' 0-8-0 heavy freight locomotive, number 49618, taken in April 1959. She was built at Crewe in April 1931 and worked for thirty years before being withdrawn for scrap in October 1961.

It is the summer of 1957 as this Fowler 'Class 7F' locomotive, number 49648, looking tired and dirty, is pictured chugging along through the rural countryside of West Yorkshire with a coal train. The coal she is burning is likely to be of poor quality as it is sending a plume of dirty black smoke high into the atmosphere. Sadly, her days were numbered and she was withdrawn from service later that year for scrap.

The design of the Fowler 'Class 7F' 0-8-0 heavy freight locomotive, as shown in the previous photograph, was in fact based on this LMS 'Class G2A' engine which had been designed by Charles Bowen-Cooke for the LNWR some thirty years earlier.

This 'Class G2A' locomotive, number 48930, is pictured at Bescot Shed near Walsall where she was based. Her shed plate, bearing the number 3A (Bescot), is clearly visible on the smokebox door at the front of the engine.

Number 48930 was built in 1903 as an LNWR 'Class G1' locomotive. She was rebuilt into a 'Class G2A' engine in 1943 and scrapped in 1963 after working for almost sixty years.

The most powerful steam locomotive ever to be built in Britain was this LNER 2-8-8-2T 'Class U1' locomotive, number 69999. Built in 1925, this 'one-off' 'Beyer-Garratt' type locomotive was designed by Nigel Gresley and the Beyer, Peacock Company who built her.

The locomotive was designed specifically as a 'banking engine' to assist heavy coal trains ascending the second steepest main line railway incline in Britain, the notorious Worsborough incline near Barnsley in South Yorkshire.

She was based at Mexborough Shed but worked at Worsborough until the 1950s when she was used by engineers working on the Manchester to Sheffield, Woodhead Line electrification scheme. She also carried out banking trials on the Lickey incline near Birmingham but was considered to be unsuitable for the job.

This very interesting photograph was taken on 11 October 1953 and shows number 69999 at Dewsnap near Manchester, being used during the heavy engineering work being carried out on the Woodhead line between Manchester and Sheffield. She was scrapped in 1956.

A dull day as this dingy LNER 'Class D15/2' locomotive, number 62503, prepares to replace another engine which has just arrived in Cambridge with '*The Fenman*' express passenger train service from Liverpool Street on 1 August 1950. Number 62503 will haul the train on the last leg of the journey to King's Lynn and Hunstanton.

An LNER 'Class B1' locomotive usually worked this daily service but this old former GER 'Class D56', built in 1900, had been chosen to work the express this particular day. Although on her last legs, she was still quite capable of completing the task, albeit in her last few months of service. She was withdrawn for scrap in February 1951.

An **LNER 4-4-0** 'Class D20/1' light passenger locomotive, number 62386, pictured approaching Staddlethorpe Station (now called Gilberdyke Station) in East Yorkshire with a Hull to Leeds (via Selby) stopping train on 13 March 1953. The locomotive, a former NER 'Class R' built in 1907, was based at Selby for her entire BR service until being withdrawn for scrap in 1956.

A cattle truck conveying live cattle is clearly visible in front of the passenger coaches on the train. This was quite common practice at the time and cattle trucks were invariably positioned immediately behind the locomotive tender, as in this picture.

LNER 'Class D41' locomotive number 2255 is pictured performing shunting duties at Elgin in Moray, Scotland on 2 July 1947, just six months before nationalisation of the railways. The letters 'L' and 'R' are missing from the LNER lettering which would have previously been displayed on her tender. This was in keeping with economy measures imposed during the war years.

Scotland remains the focus for this picture as LNER 'Class D49/1' 'Shire Class' locomotive number 62706 *Forfarshire* is pictured passing Princes Street Gardens as she hauls a local passenger train into the City of Edinburgh in 1949.

The engine, which was based locally at Edinburgh Haymarket Shed, is painted in early BR livery, with the words 'BRITISH RAILWAYS' painted on the sides of the tender, before the first BR 'Lion on Wheel' emblem was introduced.

A lone LMS 'Class 3F' locomotive, number 52154, stands quietly in some sidings at Pontefract Monkhill before picking up a local goods train in May 1957. This former Lancashire and Yorkshire Railway 'Class 27' standard goods engine was built at Horwich Works in the Victorian era, way back in 1892. Remarkably, she carried on working until 1960 when she was withdrawn from service after working almost 68 years, a testament to her design.

Another LMS 'Class 3F', former LYR 'Class 27', goods engine pictured at Pontefract Monkhill on 7 July 1958. On this occasion, number 52305 is just passing through, discharging smoke as she trundles along with a local goods train from Goole to Wakefield. Just like the 'Class 3F' engine number 52154 in the last picture, she too had a long working life, spanning some sixty-five years.

Big Bertha, a massive LMS 0-10-0 un-classified locomotive number 58100, pictured outside Derby Works on 26 August 1956, after being withdrawn from service for scrap.

Although never officially named, this 'one-off' well-known engine was referred to by all and sundry as *Big Bertha* and spent almost her entire life working as a banking engine on the Lickey incline South of Birmingham. A large spotlight, a trademark of the locomotive, had been fitted to the top front of the engine for night time working but was removed prior to this photograph being taken, although the retaining bracket is still visible.

LMS & LNER STEAM LOCOMOTIVES: THE POST WAR ERA • 95

A holiday express special excursion train from Bradford, heading to the east coastal holiday resorts of either Scarborough or Whitby, is caught on camera at Ossett in West Yorkshire in the summer of 1957. The train is a double-header, being piloted by an LNER 'Class J6' locomotive, number 64170. She is being assisted by an unidentified 'Class B1'.

Another LNER 'Class J6' locomotive is pictured standing outside Doncaster Works, circa 1947/48. She is in her LNER livery and displaying her LNER number 4203 (later BR 64203).

'Class J6' locomotives were originally built for the GNR as 'Class 521' goods engines and were first introduced in 1911. Between 1912 and 1922, further examples were built but slightly modified by Gresley and designated as 'Class 536' engines. After 1923, both classes were combined and classified as LNER 'Class J6' freight engines. Their numbers totalled 110.

BR later re-classified them as mixed traffic locomotives (2P3F) and they were frequently used to work passenger train services during the 1950s, as shown in the previous picture.

A tranquil setting in the small market town of Mildenhall, West Suffolk, is the focus for this photograph taken in the summer of 1958, as an LNER 'Class E4' locomotive, number 62785, waits patiently to depart with a two-coach, local passenger service to Cambridge. Sadly, this picturesque branch line from Mildenhall to Cambridge closed completely in 1964.

This former GER 'Class T26' mixed traffic engine (1MT), was built at Stratford Works in 1895 and continued working until being withdrawn from service in December 1959. Fortunately, however, she was preserved by the National Railway Museum in York and is currently on loan to Bressingham Steam Museum near Diss in Norfolk.

Another LNER 'Class E4' locomotive, number 62793, is pictured working a local passenger service through Thetford, Norfolk on 3 June 1952.

The train has been halted outside Thetford Signal Box and the signalman is standing alongside the locomotive giving some verbal instructions to the engine driver, who is leaning out of the cab window.

LMS 'Class 3F' locomotive number 52154 makes an un-scheduled stop whilst hauling wagons and vans through Hensall Junction in North Yorkshire. The train in question is the 11am Knottingley to Goole pick-up goods train.

Although stationary, this old former LYR 'Class 27' engine is still billowing plenty of black smoke through her chimney, adding some charm to the photograph.

LMS 'Class 7F' locomotive, number 52857, pictured at Sowerby Bridge in the Upper Calder Valley, West Yorkshire circa 1950. She is displaying her BR numbers but the tender still shows her old LMS lettering on the sides. She never did receive her full BR livery; instead, she was withdrawn from service for scrap in December 1951.

A total of 115 of these former LYR 'Class 31' engines were built new between 1912 and 1920. A further forty were rebuilt from other classes during the same period, taking the grand total to 155. Sadly, they were all scrapped.

Newly painted in her gleaming BR black livery, this LMS 'Class 5' ('Black Five') locomotive, number 45069, is pictured outside Crewe Works on 4 April 1948.

The letter 'M' prefix added to her smokebox number plate indicates Midland Region of BR. The letter 'M' is also painted on each side of her cab, above the numbers.

Prefix letters were applied to locomotives for just the first three months of 1948, to show in which of the big-four companies they had served: 'E'= LNER, 'M' = LMS, 'S'= SR, 'W'= GWR.

This was a temporary measure which was discontinued after the official BR numbering scheme was introduced in March 1948.

A well turned out LMS 'Patriot Class' locomotive, number 45509 *The Derbyshire Yeomanry*, pictured at Manchester Victoria Station with a passenger express in the late 1950s.

L M S & L N E R STEAM LOCOMOTIVES: THE POST WAR ERA • 99

This un-named 'Patriot' Class locomotive, number 45517, is pictured effortlessly working a York to Manchester express passenger train service through Castleford, West Yorkshire on 12 September 1959. The train was routed via Normanton, Wakefield and the Calder Valley Line to Manchester.

This photograph was taken late in 1947, the last year in which to view the entire fleet of LNER locomotives in their original LNER livery. The locomotive in this picture is a 'Class V2' express mixed-traffic locomotive, LNER number 888, stabled at Darlington having been temporarily taken out of service.

Upon her return to service, number 888 left Darlington Works in her new BR livery, sporting her BR number, 60888, before heading north, across the border into Scotland, then on to Aberdeen, from where she worked for the remainder of her service. She was eventually withdrawn in December 1962 before being scrapped.

LNER 'Class D49/2' 4-4-0 locomotive number 62726 *The Meynell* is pictured passing through Seamer Junction, North Yorkshire, heading towards York with an express passenger train from Scarborough in the mid-1950s.

Number 62726 was built at Darlington in 1929 and given the name *Leicestershire*, after the county by that name. In June 1932, her name was changed to *The Meynell* (named after the Meynell and South Staffordshire fox hunt), which she retained until she was withdrawn from service for scrap in December 1957.

Another photograph, this time a side view, of LNER 'Class D49/2' (Hunt Class) locomotive, number 62762 *The Fernie* (named after a fox hunt at Market Harborough in Leicestershire). This early BR photograph, showing her in steam as a light engine, was taken at York on 21 July 1951. All the 'Hunt Class' locomotives had a cast brass fox figure embellishment attached to the top of the nameplate, as can be seen in this picture.

Doncaster visits York. LNER 'Class A3' Pacific' locomotive number 60048 *Doncaster* is pictured standing in York Station with an express passenger train in April 1957.

Both the town of Doncaster and the City of York are steeped in railway history, particularly in relation to the LNER. On this occasion however, the name *Doncaster* bestowed upon the locomotive in this picture was not intended to name the engine after the famous railway town. Instead, like the names given to numerous other 'Class A3' locomotives, number 60048 *Doncaster* was named after a once famous racehorse who won the Epsom Derby in 1873. Coincidentally, number 60048 was both built and scrapped at Doncaster Works.

Peterborough in the 1950s is the setting for this picture, as LNER 'Class A3' locomotive number 60078 *Night Hawk* snakes her way through a maze of tracks and rolling stock towards her destination with an express passenger train service.

LNER 'Class A3' locomotive number 60108 *Gay Crusader* coasts her way through Finsbury Park Station, North London on the East Coast Main Line, with a down express passenger service out of King's Cross in the mid-1950s.

A man is standing on the opposite platform with (presumably) his young son in his arms, showing him the delights of watching a steam hauled train in action. Both seem captivated by the event.

LMS Ivatt 'Class 2F' 2-6-0 locomotive number 46481 pictured at Penrith Station, Cumbria, preparing to work a local branch line service to Cockermouth in the 1950s.

The branch line was recommended for closure by the Beeching Report of 1963 and part of the line from Keswick to Cockermouth closed to passengers in 1966. The track-bed from Cockermouth to Keswick formed the foundations for the building of the A66 trunk road and the remaining section of the line from Keswick to Penrith closed completely in 1972.

LMS 'Jubilee Class' locomotive number 45660 *Rooke* is pictured outside Crewe North Shed on 8 May 1948, in immaculate condition, freshly painted in her new British Railways black lined livery, in readiness to travel to Leeds Holbeck Shed (20A), where she was based for her entire BR service. She was withdrawn for scrap in July 1966.

LMS 'Royal Scot Class' locomotive number 46141 *The North Staffordshire Regiment* passes through Nuneaton Station with an up express passenger train on 23 April 1950.

LNER 'Class K3/2', number 61897 pictured working a Brighouse to Goole, Class 'H' goods train service through the cold landscape of the Yorkshire countryside on 27 February 1960.

LNER 'Class K3' locomotives were first designed by Nigel Gresley for the GNR and appeared on the scene in 1920 as GNR 'Class H4' mixed traffic engines (6MT). Just ten were built for the GNR but Gresley continued the building programme into the LNER era until a total of 193 had been reached when production ceased in 1937. The ten original GNR engines were classified by the LNER as 'Class K3/1' engines. The 183 engines built for the LNER between 1923 and 1937 were classified as 'Class K3/2' engines.

Another 'Class K3/2' mixed traffic locomotive, number 61899 this time, working an express freight train through Pontefract, heading south towards Sheffield in the autumn of 1959.

LNER 'Class B1' locomotive number 61323 pictured at Pontefract in August 1959, whilst working a special express excursion train from Hull to Southport, carrying passengers to attend the Southport Flower Show, which was the biggest flower show in the country.

The four-day event started in 1924 and has been held annually ever since. Throughout the 1950s, special excursion trains ran from various locations to convey passengers to the show, which was held in Victoria Gardens, Southport. The special train in this picture ran from Hull each year, for the four days of the event.

LMS 'Jubilee Class' locomotive number 45623 *Palestine* gathers speed as she passes through Willesden in North West London, after leaving Euston with an express passenger train service bound for Blackpool, on 20 September 1948. The engine is displaying her new BR numbers but the old LMS inscription is still being exhibited on the tender.

Euston Station in August 1955 is the setting for this photograph, as LMS 'Princess Royal Class' locomotive number 46208 *Princess Helena Victoria* reverses out of the station with '*The Shamrock*', express passenger train, which she had arrived with a short time earlier. A station pilot engine at the rear of the train would lead the coaches out of the station.

'*The Shamrock*' was an express passenger boat train service that operated daily from London Euston to Liverpool Lime Street, conveying passengers to the Irish Ferry services. The train operated from 1954 until 1966.

LMS 'Class 8F' heavy freight locomotive number 48540 pictured near Gowdall Junction, East Yorkshire, on the former Hull and Barnsley Railway line whilst working a short goods train from Hull to Cudworth near Barnsley.

The railway embankment on the extreme right of the picture carried a freight-only branch line which left the H&B Railway at Gowdall Junction and went to Pollington and Bullcroft Colliery. The day this photograph was taken, 18 October 1958, saw the last working on the Bullcroft Branch Line before it officially closed.

LMS Beyer-Garratt type locomotive number 47981 pictured in steam outside Canklow Engine Shed, Rotherham, South Yorkshire on 30 April 1950.

This 2-6-6-2T (or 2-6-6-0 + 0-6-6-2) tank locomotive, designed for heavy freight work by Henry Fowler and built by Beyer, Peacock Limited in 1930, was later fitted with a revolving coal bunker. She was withdrawn from service for scrap in 1956. A total of thirty-three were built.

Another LMS Beyer-Garratt type locomotive, number 47997, is pictured outside Crewe Works in the early 1950s, looking abandoned and neglected whilst temporarily out of service. These were never considered very successful locomotives as they were extremely heavy on coal and prone to breakdowns resulting from hot axle boxes. LMS standard axle boxes were fitted to these locomotives in accordance with Fowler's original design. However, designers from the Beyer, Peacock Company, who built the locomotives, considered that the LMS boxes were not robust enough for these engines, due to their excessive weight, and they should be replaced by sturdier boxes. The LMS disagreed and insisted that the LMS axle boxes were quite adequate so they remained in use.

LNER 4-6-2 'Class A2/3' locomotive, early BR number E500, *Edward Thompson*, pictured as a light engine at Beeston near Leeds in May 1948. The engine is turned out in early BR livery, displaying the prefix letter 'E' (Eastern Region) before its numbers. This prefix numbering system was only used for a three-month period at the start of 1948, until the official BR numbering scheme was introduced in March of that year.

Just fifteen of these locomotives were built during 1946 and 1947. Number E500 (later BR 60500) in this photograph was the first member of the class to be built and was named after the designer of this locomotive, Edward Thompson. The remaining fourteen were all named after racehorse winners of flat races. This engine was withdrawn from service in June 1963.

Another LNER 'Class A2/3' locomotive is number 60516 *Hycilla*, pictured passing Geldard Signal Box whilst traversing Geldard Curve between Leeds Central and Holbeck Low Level Stations in May 1952.

The locomotive is working the down '*Queen of Scots*' Pullman express passenger train service from King's Cross to Edinburgh Waverley and Glasgow Queen Street. The service ran continuously from 1928 (after succeeding the '*Harrogate Pullman*') until 1964, when the service was discontinued and replaced by the '*White Rose Pullman*'.

LNER 'Class B1' number 61306, pictured at Scarborough on 5 September 1959, preparing to work the 'Scarborough Flyer' express passenger train service. Number 61306, built in 1948, was withdrawn from service in 1967 and subsequently preserved. She is now working on the North Norfolk Railway. She was given the name *Mayflower* during preservation.

The *Scarborough Flyer* was first introduced in 1927 as an express passenger train which ran from London King's Cross to Scarborough Central, in both directions, during the summer months only. With the exception of disruptions during the war years, the train operated until 1963. The route from London was on the East Coast Main Line to York, then to Scarborough via Malton.

This fine picture shows LNER 'Class B1' mixed traffic locomotive number 61291 working hard as she hauls an up express passenger train past King Edward Bridge Junction signal box, Newcastle, in the 1950s.

A total of 410 of these fine locomotives, designed by Edward Thompson, were built over a ten-year period between 1942 and 1952. Just two have been preserved, one of which (number 61306) appears in the previous photograph.

LNER 'Class B1' number 61159 pictured slowly meandering through the village station of Golcar near Huddersfield, whilst working the 10.25am Manchester London Road to London Marylebone express passenger service on 14 May 1950. This train was being diverted via Huddersfield, due to the temporary closure of the Woodhead Tunnel for track maintenance work to be carried out.

A nice display of smoke emanating from this LNER 'Class B1' locomotive, number 61145, as she races through Paddock on the outskirts of Huddersfield with the 9.50am London Marylebone to Manchester London Road (re-named Manchester Piccadilly in 1960) express passenger train on Sunday 14 May 1950.

This picture was taken on the same day as the previous photograph but the train is travelling in the opposite direction and at a different location. As in the previous picture, this train was also being diverted, due to Sunday track maintenance work causing the temporary closure of a part of the Woodhead line between Sheffield and Manchester.

LOCOMOTIVE TECHNICAL DETAILS AND GENERAL INFORMATION

Locomotive Wheel Configuration

Wheel configuration or wheel arrangement describes the way in which the distribution of locomotive wheels (wheel assembly) is counted. Several different methods have been used to identify these configurations worldwide but the method adopted in Britain and Ireland (also the USA and Canada) for use on steam locomotives was the 'Whyte System' or 'Whyte Notation'.

The system consisted of three digits, separated by hyphens, e.g. 4-6-0, 2-6-2, 0-8-0. The three sets of numbers represent the order in which the wheels are arranged, starting at the front of the locomotive. They represent the number of leading wheels, the number of coupled driving wheels and the number of trailing wheels. Tender wheels are not included.

Tank engines did not have a tender for carrying water and coal. Coal was carried in a bunker behind the cab and water was carried in various styles of tanks. The most common being side tanks which were fitted to each side of the boiler (ordinary / side tank engine). Other types most commonly used were saddle tanks, pannier tanks and well tanks. The wheel configuration number also revealed if the locomotive was a tank engine, and the type of tank. This was indicated by using a suffix, e.g. 2-6-4T (ordinary tank engine), 0-6-0PT (pannier tank engine), 0-6-0ST (saddle tank engine), etc.

Whyte Notation – Wheel Assembly Names

Most wheel arrangements under the Whyte system were given names. Referring to a locomotive by its wheel arrangement name is common in some countries, particularly in America, where many of the names apply solely to American locomotives. People may be surprised to learn that there are in excess of one-hundred names which refer to various locomotive wheel arrangements worldwide. In Britain, most of the names are seldom, if ever, used.

Just a handful have been used in Britain over the years, the main one being a 'Pacific' locomotive. Many of the big powerful passenger steam locomotives in Britain have been referred to over the years as 'Pacifics', and some people are unaware where the name originates from. It is in fact a name listed in the Whyte Notation Chart as being a locomotive with a 4-6-2 wheel arrangement. I have selected just four names, which are the only names generally used from the Whyte Classification Chart when referring to British locomotives. They are listed as follows:

4-6-2 locomotive is a 'Pacific'.
4-4-2 locomotive is an 'Atlantic'.
2-6-0 locomotive is a 'Mogul'.
2-6-2 locomotive is a 'Prairie'.

The former Great Western Railway built a number of large 2-6-2 tank engines and some smaller 2-6-2 tank engines which they officially classified as 'Large Prairie' and 'Small Prairie'.

DETAILS OF LOCOMOTIVES FEATURED IN THIS BOOK

(In numerical order)

LONDON, MIDLAND AND SCOTTISH RAILWAY (LMS) LOCOMOTIVES

LMS No. 8 (Railway Carriage Department Number) *Earlestown*

This LMS (LNWR) 0-6-0 Saddle Tank 'Special Tank Class' locomotive was built by Francis Webb at Crewe in August 1878 for the London and North Western Railway Company as LNWR 'Special Tank' locomotive number 2359. Absorbed into LMS stock in 1923, she became an LMS departmental shunting locomotive at Earlestown Works in Newton-le-Willows, Lancashire (now Merseyside). She was named *Earlestown* after the works and allocated LMS Carriage Departmental number CD8. She was later transferred to Wolverton Works in Buckinghamshire before being absorbed into the BR locomotive stock in 1948. She was not allocated a new BR number but continued using her old LMS departmental number 8. She carried on performing shunting duties at Wolverton Works until she was withdrawn from service for scrap in October 1957. (Photo P.24)

LMS (LNWR) 0-6-0ST 'Special Tank' (Saddle Tank) Locomotives: general information

A total of 260 'Special Tank' locomotives were built by Francis William Webb for the LNWR between 1870 and 1880. Although designed as shunting engines (power classification 2F), they were primarily a saddle tank version of the highly successful LNWR 'DX Class' goods tender locomotive, of which 943 were designed and built by John Ramsbottom between 1858 and 1862.

Withdrawals of the 'Special Tank' locomotives started in 1920 but 250 were absorbed into the LMS Railway in 1923. Six members of the class were assigned to Wolverton Works as departmental shunting locomotives, numbers CD1, CD2, CD3, CD5, CD6 and CD7 (Carriage Department). A seventh departmental engine, number CD4, was assigned to shunting duties at Crewe Works along with LNWR number 3323, which retained her original LNWR number.

LNWR number 2359 was transferred to Earlestown depot in Lancashire where she became a works departmental shunting locomotive number CD8 and was given the name *Earlestown* after the depot where she worked. Number CD8 was subsequently transferred from Earlestown Works to Wolverton Works. She retained the name *Earlestown* and her CD8 departmental number.

The vast majority of 'Special Tanks' had been withdrawn from service and scrapped by the outbreak of the Second World War in 1939 and only five survived into the BR era. All five were Wolverton departmental locomotives and they continued working there until the last three were withdrawn from service in 1959.

Other LMS locomotives pictured in this book displaying their old LMS numbers are set out below. Information about these engines is listed under their respective BR numbers:

No. 2219 – See BR number 42219
No. 2226 – See BR number 42226
No. 5530 – See BR number 45530
No. 7406 – See BR number 47406
No. 7892 – See BR number 47892
No. 8921 – See BR number 48921
No. 12435 – See BR number 52435

The remaining locomotives pictured in this book are all displaying their BR numbers, under which all details and technical information about each locomotive can be found.

LMS (MR) 4-4-0 'Class 2P' locomotives - general information

Throughout their history, the Midland Railway Company had a 'small engine policy', to build and use small locomotives as opposed to large locomotives to operate both passenger and freight train services. 'Class 2P' locomotives were a prime example.

LMS 'Class 2P' 4-4-0 locomotives were originally designed by Samuel Johnson as a development of the Midland Railway 'Class 483' locomotive, and a total of 275 were built between 1876 and 1901 in a series of twelve different classes.

The twelve different classes, which were later combined and re-classified as Midland Railway 'Class 2' (LMS 'Class 2P') locomotives, are listed as follows:

'Class 60' (forty built, 1898–1899)
'Class 156' (twenty built, 1896–1901)
'Class 1312' (ten built, 1876)
'Class 1327' (twenty built, 1876)
'Class 1562' (thirty built, 1882–1884)
'Class 1667' (ten built, 1884)
'Class 1738' (twenty built, 1885)
'Class 1808' (twenty-five built, 1888–1891)
'Class 2183' (twenty-five built, 1892–1896)
'Class 2203' (forty-five built, 1893–1895)
'Class 2421' (twenty built, 1899)
'Class 2581' (ten built, 1900)

All twelve classes listed above, which later combined to become LMS 'Class 2P' locomotives, were originally designed and built for use as express passenger locomotives until they were replaced by more powerful engines, after which they were used to haul local and secondary passenger train services (light passenger work). All the 4-4-0 '2P' locomotives designed by Samuel Johnson were rebuilt, some several times, during their lifespan. The practice of re-building existing locomotives as well as building new ones continued under the auspices of Richard Deeley after the retirement of Johnson in 1901, and from 1909 under his successor, Henry Fowler, who went on to become Chief Mechanical Engineer for the LMS Railway Company. In total, seven hundred 'Class 2P' locomotives were built for the MR and LMS railway until 1932 when the building programme ceased.

When the railways were nationalised, a total of 136 'Class 2P' locomotives were still in service and were transferred into the stock of BR locomotives in 1948. With the exception of two, they all survived until the end of the 1950s before being withdrawn for scrap. The last member of the class survived until 1962. Due to their small size they were frequently used as 'double-headers' to work passenger trains.

No. 40538. LMS 4-4-0 'Class 2P' locomotive. Designed by Samuel Johnson and built at Derby in September 1899 as a Midland Railway 'Class 60' (later MR 'Class 2') locomotive, number 538. Absorbed into LMS in 1923 and re-classified as LMS 'Class 2P' number 538. Allocated BR number 40538 in 1948. Withdrawn from service in May 1959, whilst based at Derby. She was stored in Chaddesden Sidings, Derby before being scrapped by Albert Looms Ltd, Derby in October 1960. (Photo P.83)

No. 40556. LMS 4-4-0 'Class 2P' locomotive. Designed by Samuel Johnson and built at Glasgow in May 1901 by Neilson & Co. as a Midland Railway 'Class 60' (later MR 'Class 2') locomotive number 556. Absorbed into LMS in 1923 and re-classified as LMS 'Class 2P' number 556. Allocated BR number 40556 in 1949. Withdrawn from service in April 1956, whilst based at Hasland Shed, Chesterfield, Derbyshire and subsequently scrapped in Derby Works. (Photo P.73)

No. 40581. LMS 4-4-0 'Class 2P' locomotive. Designed by Henry Fowler and built at Derby Works in July 1928 as LMS 'Class 2P' locomotive number 581 (BR number 40581 from 1948). Withdrawn from service in October 1960, whilst based at Royston Shed, South Yorkshire. Scrapped at Doncaster BR Works a few days later. (Photo P.72)

No. 40602. LMS 4-4-0 'Class 2P' locomotive. Designed by Henry Fowler and built at Derby Works in November 1928 as LMS 'Class 2P' locomotive number 602 (BR number 40602 from 1948). Withdrawn from service in October 1961, whilst based at Corkerhill Shed, Glasgow and later scrapped at Connells Scrapyard, (Coatbridge Yard), Glasgow. (Photo P.73)

No. 40613. LMS 4-4-0 'Class 2P' locomotive. Designed by Henry Fowler and built at Derby Works in October 1929 as LMS 'Class 2P' locomotive number 613 (BR number 40613 from 1948). Withdrawn from service in October 1961, whilst based at Corkerhill Shed, Glasgow. Scrapped in June 1963 at Connells Scrapyard (Calder Yard), Glasgow. (Photo P.83)

No. 40630. LMS 4-4-0 'Class 2P' locomotive. Designed by Henry Fowler and built at Derby Works in January 1930 as LMS 'Class 2P' locomotive number 630 (BR number 40630 from 1948). Withdrawn from service in October 1960 whilst based at Normanton Shed (West Yorkshire). Scrapped at Doncaster BR Works in November 1960. (Photo P.15 and on the front cover.)

LMS 4-4-0 'Class 4P' Fowler compound locomotives: general information

These 3-cylinder compound locomotives, designed by Henry Fowler, were based on the Midland Railway 'Class 1000' locomotive, designed and built by Samuel Waite Johnson. A total of forty-five 'Class 1000' engines were built at Derby between 1902 and 1909. The whole class were later rebuilt between 1913 and 1928.

Henry Fowler built a total of 195 'Class 4P' compound locomotives between 1924 and 1932 to add to the forty-five 'Class 1000' engines already in service. Apart from a few minor alterations, the 'Class 4P' and the 'Class 1000' were identical in design. The main difference between the two classes of locomotives could be found in the driving wheels. The 'Class 1000' had 7ft (2.134m) wheels and the Fowler 4P design had 6ft 9in (2.057m) driving wheels.

The LMS 'Class 4P' locomotives were designed for use as passenger locomotives (power classification 4P) and were extremely successful. In 1928, number 1054 (built in 1924) became the first locomotive ever to travel non-stop from London to Edinburgh, a distance of about 400 miles (640km), when she hauled the LMS *'Royal Scot'* express passenger service from Euston via the West Coast Main Line. This was four days before the LNER *'Flying*

Scotsman' made her first non-stop run from London King's Cross to Edinburgh on the East Coast Main Line. The withdrawal from service of 'Class 4P' locomotives began in 1952 and continued into the 1960s. The last two were withdrawn in 1961. Only the first member of the original 'Class 1000', Midland Railway number 1000, was preserved and is on public display in the National Railway Museum in York. None of Fowler's 'Class 4P' compounds have been preserved.

No. 40931. LMS 4-4-0 'Class 4P' (Fowler Compound) locomotive. Built at the Vulcan Foundry, Newton-le-Willows, Lancashire in May 1927 as LMS 'Class 4P' locomotive number 931 (BR number 40931 from 1948). Withdrawn from service in October 1958, whilst based at Lancaster Green Ayre Shed. Scrapped in June 1959 by Bush & Sons, Pye Bridge, Alfreton, Derbyshire. (Photo P.46)

No. 40937. LMS 4-4-0 'Class 4P' (Fowler Compound) locomotive. Built at Derby Works in September 1932 as LMS 'Class 4P' locomotive number 937. Withdrawn from service in April 1958 whilst based at Lancaster Green Ayre Shed. Scrapped in May 1958 at Crewe BR Works. (Photo P.46)

No.41105. LMS 4-4-0 'Class 4P' (Fowler Compound) locomotive. Built at Derby Works in November 1925 as LMS 'Class 4P' locomotive number 1105. Withdrawn from service in September 1957 and scrapped shortly afterwards. (Photo P.88)

No. 41151. LMS 4-4-0 'Class 4P' (Fowler Compound) locomotive. Built by the North British Locomotive Company, Glasgow in August 1925 as LMS 'Class 4P' locomotive number 1151. Withdrawn from service in January 1957 whilst based at Lancaster Green Ayre Shed. Scrapped at Derby BR Works shortly after being withdrawn from service. (Photo P.37)

No. 41152. LMS 4-4-0 'Class 4P' (Fowler Compound) locomotive. Built by the North British Locomotive Company, Glasgow in September 1925 as LMS 'Class 4P' locomotive number 1152. Withdrawn from service in March 1958 whilst based at Lancaster Green Ayre Shed. Scrapped at Derby BR Works shortly after being withdrawn from service. (Photo P.88)

No. 41154. LMS 4-4-0 'Class 4P' (Fowler Compound) locomotive. Built by the North British Locomotive Company, Glasgow in September 1925 as LMS 'Class 4P' locomotive number 1154. Withdrawn from service in August 1955 whilst based at Trafford Park Shed, Manchester. Scrapped at Derby BR Works shortly after being withdrawn from service. (Photo P.37)

No. 41159. LMS 4-4-0 'Class 4P' (Fowler Compound) locomotive. Built by the North British Locomotive Company, Glasgow in October 1925 as LMS 'Class 4P' locomotive number 1159. Withdrawn from service in April 1958 and scrapped in September 1958. (Photo P.12)

LMS 2-6-2T 'Class 2P' Ivatt tank locomotives: general information
A total of 130 of these 2-6-2 tank locomotives, designed by Henry George Ivatt (known as George Ivatt), were built at Crewe and Derby between 1946 and 1952. They were designed for light passenger work and given an LMS power classification 2P. When the engines were taken into BR stock it was decided to use them as mixed traffic locomotives and they were re-classified as BR power classification 2MT. Due to their size, they were given the nickname 'Mickey Mouse'. The locomotive design formed the basis of the 'Standard Class 2', 2-6-2 tank locomotive which was later produced by BR.

No. 41204. LMS 2-6-2T 'Class 2P' (Ivatt) locomotive. Built at Crewe Works in December 1946 as LMS 'Class 2P' locomotive number 1204. Withdrawn from service in December 1966 whilst based at Stockport Edgeley Shed. Scrapped by T. W. Ward, Killamarsh, Derbyshire in April 1967. (Photo P.69)

No. 41206. LMS 2-6-2T 'Class 2P' (Ivatt) locomotive. Built at Crewe Works in December 1946 as LMS 'Class 2P' (re-classified '2MT' by BR) locomotive number 1206. Withdrawn from service in March 1966 and scrapped in July 1966. (Photo P.84)

No. 41251. LMS 2-6-2T 'Class 2P' Ivatt design locomotive was built for BR at Crewe Works in November 1949 as BR 'Class 2MT' locomotive number 41251. Withdrawn from service in March 1966 and scrapped in July 1966. (Photo P.79)

No. 41253. LMS 2-6-2T 'Class 2P' Ivatt design locomotive was built for BR at Crewe Works in November 1949 as BR 'Class 2MT' locomotive number 41253. Withdrawn from service in April 1964 whilst based at Lancaster Green Ayre, and remained there until she was transported to Crewe BR Works where she was cut up for scrap in August 1964. (Photos P.32, 69 & 70)

No. 41278. LMS 2-6-2T 'Class 2P' Ivatt design locomotive was built for BR at Crewe Works in October 1950 as BR 'Class 2MT' locomotive number 41278. Withdrawn from service in November 1962 and scrapped in February 1963. (Photo P.80)

LMS 0-4-4T 'Class 2P' Stanier locomotives: general information

Just ten of these 0-4-4 tank locomotives were built at Derby between 1932 and 1933. They were the first LMS design of the then, newly appointed, Chief Mechanical Engineer, William Stanier. Originally, they were all fitted with 'stove-pipe' chimneys which were later replaced by Stanier chimneys after he had perfected the design. Two members of the class (BR numbers 41908 and 41909) were fitted with push-pull apparatus in 1934 for use on the St Albans branch line. As a result of their success, all remaining class members were fitted with the apparatus by BR during the 1950s.

The whole class was withdrawn from service for scrap in 1959, with the exception of BR number 41900, the first member of the class to be built, and which features in this book (P.84). She survived until March 1962 before being withdrawn from service and scrapped.

No. 41900. LMS 0-4-4T 'Class 2P' (Stanier) design locomotive was built at Derby Works in December 1932 as LMS number 6400, later re-numbered LMS 1900, before becoming BR 41900 after 1948. Withdrawn from service in March 1962 and scrapped in April 1962. (Photo P.84)

LMS 2-6-4T 'Class 4P' Fairburn tank locomotives: general information

A total of 277, 2-6-4 tank locomotives, designed for the LMS Railway by Charles E. Fairburn, were built for both the LMS and BR between 1945 and 1951. The design was based on an earlier Stanier, 2-6-4 tank engine (see locomotive number 42650 below), but with a shorter wheelbase and some other modifications. The Stanier 2-6-4 tank itself had been derived from an earlier successful 4P, 2-6-4 tank engine, which had been designed and built by Henry Fowler in 1927 (see locomotive number 42333 below).

The Fairburn 4P design was re-classified as BR 'Class 4MT' locomotive in 1948. The later BR Standard 'Class 4' tank engine was based on this Fairburn-designed locomotive.

No. 42219. LMS 2-6-4 'Class 4P' Fairburn-designed tank locomotive. Built at Derby Works in January 1946 as LMS number 2219. Re-numbered BR 42219 in October 1948. Withdrawn from service in June 1962 whilst based at Shoeburyness Shed, and scrapped at Doncaster Works in October 1962. (Photo P.14)

No. 42226. LMS 2-6-4 'Class 4P' Fairburn design tank locomotive. Built at Derby Works in May 1946 as LMS number 2226. Re-numbered BR 42226 in March 1949. Withdrawn from service in June 1964, whilst based at Stoke Shed and scrapped at Cashmores, Great Bridge, Tipton, West Midlands in May 1964. (Photo P.31)

No. 42285. LMS 2-6-4 'Class 4P' Fairburn design tank locomotive. Built at Derby Works in September 1947 as LMS number 2285. Re-numbered BR 42285 in September 1950. Withdrawn from service in September 1965, whilst based at Low Moor Shed, Bradford, and scrapped by W. George & Son, Wath-upon-Dearne, South Yorkshire in October 1965. (Photo P.32)

LMS 2-6-4T 'Class 4P' Fowler tank locomotives: general information

A total of 125 'Class 4P' 2-6-4 tank locomotives, designed by Henry Fowler, were built at Derby Works between 1927 and 1934. They were originally designed to be used as LMS passenger locomotives (power classification 4P) but when they were taken into BR stock after 1948 they were used for mixed traffic duties and re-classified as 4MT.

These locomotives were not designed with cab side windows or cab doors, although both were fitted to the last thirty members of the class to be built (numbers 2395 to 2424), as can be seen in the photograph of number 42407 which appears on page sixteen.

Large numbers of these Fowler tank engines were used to work commuter trains in and out of large towns and cities (including London), whilst others were used on quiet rural lines such as the Central Wales Line from Swansea Victoria to Shrewsbury. They were also successfully used as banking engines. The whole class of 125 locomotives was withdrawn from service between 1959 and 1966 and cut up for scrap. Unfortunately, none of these locomotives have been preserved.

No. 42333. LMS 2-6-4T 'Class 4P' Fowler design tank locomotive. Built in March 1929 at Derby Works as LMS number 2333. Re-classified by BR after 1948 as a 'Class 4MT' tank locomotive, number 42333. Withdrawn from service in May 1963 and scrapped later that month. (Photo P.89)

No. 42407. LMS 2-6-4T 'Class 4P' Fowler design tank locomotive. Built in October 1933 at Derby Works as LMS number 2407. Re-classified by BR as 'Class 4MT' tank locomotive, number 42407. Withdrawn from service in November 1962, whilst based at Mirfield Shed in West Yorkshire. Scrapped at Darlington BR Works in March 1964. (Photo P.16)

LMS 2-6-4T 'Class 4P' Stanier tank locomotives: general information

A total of 206 LMS 'Class 4P' 2-cylinder tank locomotives, designed by William Stanier, were introduced between 1935 and 1943. The locomotive design was based upon the very similar but more complicated design of his 3-cylinder 2-6-4 tank locomotive which he had introduced the previous year.

His 3-cylinder locomotives were specifically designed to work passenger commuter services on the London, Tilbury and Southend line and a total of thirty-seven had been introduced in 1934. The extra third cylinder provided the increased acceleration required for the frequent station stops that existed on the commuter lines.

After the railways were nationalised in 1948, both the 2- and the 3-cylinder versions of the 2-6-4 tank engines were re-classified from '4P' to '4MT' to facilitate their use as mixed traffic locomotives.

The thirty-seven 3-cylinder locomotives were all withdrawn from service between 1960 and 1962. Just one, number 42500, was preserved and the remainder were scrapped. All 206 of the 2-cylinder versions were withdrawn from service for scrap over an eight year period from 1960 to 1967. None was preserved.

No. 42650. LMS 2-6-4T 'Class 4P' Stanier design 2-cylinder, tank locomotive. Built in December 1938 at Derby Works as LMS number 2650. Later re-classified by BR as a 'Class 4MT' number 42650. Withdrawn from service in June 1967 and scrapped in January 1968. (Photo P.16)

LMS 2-6-0 'Class 5F' (Crab Class) locomotives: general information

A total of 245 of these highly successful locomotives, designed by George Hughes, were built at Crewe Works (175) and Horwich Works (70) between 1926 and 1932. They were easily recognisable by the appearance of their two highly angled outside cylinders.

They were originally designed and built as light/medium goods locomotives and they were allocated a power classification of '5F'. It was quickly established, however, that they were ideally suited for working passenger train services and their power classification was subsequently re-classified to '6P5F'. They were further re-classified as '5MT' by BR and continued working successfully as mixed traffic locomotives until the end of the steam era.

They were commonly known as 'Crab Class' locomotives or 'Crabs' but occasionally they were referred to as 'Horwich Moguls'. After working on the LMS, all 245 of these engines entered BR service in 1948 and they all survived into the 1960s. Withdrawals started in 1961 and continued until the last two (numbers 42727 and 42942) were withdrawn from service in 1967. The first member of the class, number 42700 (LMS 13000), which was built at Horwich, has been preserved and can be viewed at the National Railway Museum in York.

No. 42715. LMS 2-6-0 'Class 5F' (Crab Class) locomotive. Built at Horwich Works in April 1927 as LMS number 13015. Renumbered 2715 in 1934 and BR number 42715 in 1948. Later re-classified by BR as a '5MT' locomotive. Withdrawn from service in February 1966 whilst based at Stockport Edgeley Shed, and scrapped by Cashmores in June 1966. (Photo P.49)

No. 42944. LMS 2-6-0 'Class 5F' (Crab Class) locomotive. Built at Crewe Works in December 1932 as LMS number 13244. She was the last member of the class to be built. She was later renumbered LMS 2944 before receiving her BR number 42944 in 1948. She was later re-classified as a '5MT' locomotive by BR. Withdrawn from service in April 1963 whilst based at Birkenhead Shed. She was scrapped at Horwich Works in June 1963. (Photo P.49)

LMS 0-6-0 'Class 4F' Fowler ('Duck Six') locomotives: general information

A total of 575 of these locomotives, designed by Henry Fowler, were built between 1924 and 1941. The locomotive design was based upon Fowler's earlier Midland Railway 'Class 3835' which was introduced in 1911 and was very similar in design, apart from having a left-hand drive as opposed to a right-hand drive.

The 'Class 4F' was designed and built as a standard goods locomotive to work medium freight traffic and they earned the nickname 'Duck Sixes' due to their 0-6-0 wheel arrangement. Unfortunately, the 'Duck Sixes' frequently suffered mechanical failures due to 'hot axle boxes' caused by overheating of the bearings. This problem was later overcome by the installation of mechanical lubricators, after which they became successful and reliable locomotives. Although designed for freight duties, in later years it was not uncommon to see them working local passenger services.

No. 44082. LMS 0-6-0 'Class 4F' locomotive. Designed by Henry Fowler and built by Kerr, Stuart & Co. Ltd, Stoke-on-Trent, Staffordshire in September 1925 as LMS number 4082 (later BR 44082). Withdrawn from service in August 1961 whilst based at Canklow Shed, Rotherham and scrapped at Crewe Works in September 1961. (Photo P.35)

No. 44128. LMS 0-6-0 'Class 4F' locomotive. Designed by Henry Fowler and built at Crewe Works in August 1925 as LMS number 4128 (later BR 44128). Withdrawn from service in December 1962 whilst based at Barrow Hill (Staveley) Shed near Chesterfield, Derbyshire. Scrapped at Derby BR Works in June 1963. (Photo P.35)

LMS 4-6-0 'Class 5' ('Black Five') Stanier locomotives: general information

A total of 842 LMS 'Class 5' ('Black Five') mixed traffic locomotives, designed by William Stanier, were built between 1934 and 1951. Shortly after being introduced into service they became known as 'Black Staniers' due to being painted black as opposed to the red livery used on the Stanier 4-6-0 'Jubilee Class' locomotives being built at the same time. Later they became known as 'Black Fives' after the LMS power classification of 5P5F, which was painted on both sides of the cab.

'Black Five' locomotives were designed as an all-purpose mixed traffic locomotive and were the LMS equivalent of the highly successful GWR '4900 Class' (Hall Class) locomotives. The 'Black Fives' turned out to be the most reliable and efficient design of general purpose locomotive ever built in Britain. They were extremely popular amongst the drivers and were used all over the country. Withdrawals of the 'Class 5' locomotives from service took place between 1961 and 1968 and they were amongst the last steam locomotives to operate on the railway network before steam was replaced by diesel and electric traction. The power classification of the 'Class 5' was changed from 5P5F to 5MT by BR after nationalisation of the railways took place. A total of eighteen 'Black Fives' have been preserved.

No. 44722. LMS Design 4-6-0 'Class 5' ('Black Five') locomotive. Built by BR at Crewe in April 1949. Withdrawn from service in April 1967 whilst based at Perth South Shed in Scotland. Scrapped at Inslow Works, Motherwell in September 1967. (Photo P.60)

No. 44738. LMS Design 4-6-0 'Class 5' ('Black Five') locomotive. Built by BR at Crewe in June 1948 and put to work at Llandudno Junction Engine Shed. She was later transferred to Speke near Liverpool from where she was withdrawn from service in June 1964. She was scrapped in September 1964 at Crewe Works. This locomotive was built with experimental Caprotti valve gear which gave the locomotive a distinctive appearance. (Photo P.61)

No. 44755. LMS 4-6-0 'Class 5' ('Black Five') locomotive. Built by BR at Crewe Works in May 1948 and put to work at Derby. She was later transferred to Stockport from where she was withdrawn from service in November 1963. Scrapped at Crewe BR Works February 1964. This locomotive was fitted with a steel firebox. (Photo P.61)

No. 44757. LMS 4-6-0 'Class 5' ('Black Five') locomotive. Built by BR at Crewe Works in December 1948 and put to work at Holbeck Shed in Leeds. Withdrawn from service in November 1965 whilst based at Brunswick Shed, Liverpool. Scrapped at Cashmores Scrapyard, Great Bridge, Tipton, Staffordshire in February 1966. This locomotive was one of twenty Black Fives to be fitted with Caprotti valve gear in 1948 and one of just three to receive a double chimney. She was also fitted with a steel firebox. (Photo P.41)

No. 44854. LMS 4-6-0 'Class 5' ('Black Five') locomotive. Built at Crewe Works in December 1944 as LMS number 4854 (later BR number 44854). Worked in the Leeds and Wakefield areas of West Yorkshire until being withdrawn from service in October 1967, whilst based at Normanton Shed. Scrapped in October 1967 by T. W. Ward of Sheffield. (Photo P.18)

No. 45069. LMS 4-6-0 'Class 5' ('Black Five') locomotive. Built in January 1935 by the Vulcan Foundry at Newton-le-Willows, Lancashire, as LMS number 5069 (later BR number 45069). Withdrawn from service in June 1967 whilst based at Crewe South Shed. Scrapped in December 1967 by Cohens of Kettering. (Photo P.98)

No. 45211. LMS 4-6-0 'Class 5' ('Black Five') locomotive. Built in June 1935 by the Vulcan Foundry at Newton-le-Willows,

Lancashire, as LMS number 5211 (later BR number 45211). Withdrawn from service in May 1967 and scrapped by Drapers of Hull in October 1967. (Photo P.26)

No. 45225. LMS 4-6-0 'Class 5' ('Black Five') locomotive. Built by Armstrong Whitworth of Newcastle in August 1936 as LMS number 5225 (later BR number 45225). Withdrawn from service in October 1967 whilst based at Stockport Edgeley Shed. Scrapped by Cashmores of Newport, South Wales in July 1968. (Photo P.41)

LMS 4-6-0. Fowler 'Patriot Class' locomotives: general information

A total of fifty-two 'Patriot Class' locomotives, designed by Henry Fowler, were built for the LMS Railway at Crewe (42) and Derby (10) between 1930 and 1934. They were based on the designs of two locomotives, the 'Royal Scot Class' and the LNWR 'Claughton Class' ('large Claughton'). The chassis design was based on the 'Royal Scot' chassis and the boiler design was based on the 'Claughton Class' locomotives. Consequently, the 'Patriot' locomotives were nicknamed *Baby Scots*. The majority, but not all, 'Patriot Class' locomotives were given names.

'Patriot Class' locomotives were designed for use as express passenger locomotives with a power classification of 5XP (a classification between 5P and 6P). After nationalisation BR utilised the locomotives for mixed traffic duties and they were re-classified as 6P5F in 1951. A number of 'Patriot Class' locomotives were later rebuilt by George Ivatt into another class of locomotive called the 'Rebuilt Patriot Class'.

No. 45509 *The Derbyshire Yeomanry*. LMS 4-6-0 'Patriot Class' locomotive. Built at Crewe Works in August 1932 as LMS number 6005 (re-numbered 5509 from 1934). Withdrawn from service in August 1961 whilst based at Newton Heath Shed, Manchester and scrapped at Crewe Works in September 1961. This locomotive was given the name *The Derbyshire Yeomanry* by BR in May 1951. (Photo P.98)

No. 45517. LMS 4-6-0 'Patriot Class' locomotive. Built at Crewe in February 1933 as LMS number 5592 (re-numbered 5517 from 1934). Withdrawn from service in May 1962 whilst based at Bank Hall Shed, Birkenhead. Scrapped at Crewe BR Works in July 1962. This locomotive was unnamed. (Photo P.26 & 99)

No. 45519 *Lady Godiva*. LMS 4-6-0 'Patriot Class' locomotive. Built at Crewe in February 1933 as LMS number 6008 *Lady Godiva* (re-numbered 5519 from 1934). Withdrawn from service in March 1962, whilst based at Bristol Barrow Road Shed. Scrapped at Crewe Works later that month. (Photo P.74)

LMS IVATT 'Rebuilt Patriot Class' 4-6-0 locomotives: general information

A total of eighteen 'Patriot Class' locomotives were rebuilt between 1946 and 1949 to create a new class of locomotive called the 'Rebuilt Patriot Class'. They were rebuilt by George Ivatt with large taper boilers, new cylinders and double chimneys. They were given a 6P power classification which was later re-classified to 7P by BR.

The remaining thirty-four 'Patriot Class' locomotives which were not rebuilt were thereafter referred to as 'Un-rebuilt Patriot Class' locomotives.

No. 45521 *Rhyl*. LMS 4-6-0 'Rebuilt Patriot Class' locomotive. Built at Derby Works in March 1933 as 'Patriot Class' LMS number 5533 (re-numbered 5521 from 1934). Allocated the name *Rhyl* in 1937. Rebuilt in November 1946 as a 'Rebuilt Patriot Class' and in 1951, her power classification was upgraded by BR from 6P to 7P. She was withdrawn from service in September 1963 whilst based at Wigan Springs Branch Shed. She was scrapped at Crewe BR Works in November 1963. (Photo P.29)

No. 45530 *Sir Frank Ree*. LMS 4-6-0 'Rebuilt Patriot Class' locomotive. Built at Crewe Works in April 1933 as LMS 'Patriot Class' locomotive number 5530. Received her rebuild to a 'Rebuilt Patriot Class' in October 1946. Allocated the name *Sir Frank Ree* in March 1937. She was given BR number 45530 in April 1948 and her power classification was later upgraded by BR from 6P to 7P. Withdrawn from service in January 1965 whilst based at Carlisle Kingsmoor Shed, and scrapped at Inslow Works, Motherwell in March 1966. This locomotive was named after Sir Frank Ree who was a former Director of the London and North Western Railway and the North London Railway Companies. (Photo P.74)

LMS 4-6-0 'Jubilee Class' Stanier locomotives: general information
A total of 191 'Jubilee Class' locomotives, designed by William Stanier, were built between 1934 and 1936. The locomotives were named 'Jubilee Class' locomotives to commemorate the 'Silver Jubilee' of King George V, which took place in May 1935. The first member of the class, number 5552 (later BR 45552), which was built in June 1934, was allocated the name *Silver Jubilee* in April 1935.

The 'Jubilee Class' locomotives were built at Crewe (131), Derby (10) and the North British Works in Glasgow (50) for use as express passenger locomotives. After being built, they left the workshops painted in 'crimson lake' red livery. Their power classification was originally LMS 5XP (a power classification between 5P and 6P). They were later re-classified by BR to 6P in 1951 and revised to 6P5F in November 1955 when they started to be used as mixed traffic locomotives.

One member of the class, number 45637 *Windward Isles*, was scrapped in 1952 following an accident. The remainder were withdrawn from service between 1960 and 1967. Four members of the class have been preserved.

No. 45566 *Queensland*. LMS 4-6-0 'Jubilee Class' locomotive. Built in August 1934 at the North British Locomotive Works, Glasgow as LMS number 5566. Allocated BR number 45566 in 1948. Withdrawn from service whilst based at Leeds Holbeck Shed in November 1962 and scrapped at Crewe Works in December 1962. (Photo P.110)

No. 45568 *Western Australia*. LMS 4-6-0 'Jubilee Class' locomotive. Built in August 1934 at the North British Locomotive Works, Glasgow as LMS number 5568. Allocated BR number 45568 in 1948. Withdrawn from service in April 1964 whilst based at Newton Heath Shed, Manchester and scrapped by Drapers of Hull in January 1965. (Photo P.43)

No. 45572 *Eire*. LMS 4-6-0 'Jubilee Class' locomotive. Built by the North British Locomotive Company, Glasgow, in September 1934 as LMS number 5572. Allocated BR number 45572 in 1948. Withdrawn from service in January 1964 whilst based at Willesden Shed, North West London. Cut up by Albert Looms Ltd, Derby in July 1964. (Photo P.43)

No. 45623 *Palestine*. LMS 4-6-0 'Jubilee Class' locomotive. Built at Crewe Works in October 1934 as LMS number 5623. Allocated BR number 45623 in 1948. Withdrawn from service in July 1964 whilst based at Newton Heath Shed, Manchester and scrapped by the Central Wagon Company, Wigan in September 1964. (Photo P.105)

No. 45660 *Rooke*. LMS 4-6-0 Stanier 'Jubilee Class' locomotive. Built at Derby Works in December 1934 as LMS number 5660. Allocated BR number 45660 in 1948. Withdrawn from service in July 1966 whilst based at Leeds Holbeck Shed. Scrapped by Drapers of Hull in October 1966. (Photo P.103)

LMS 4-6-0 Fowler/Stanier 'Royal Scot Class' locomotives: general information
A total of seventy 'Royal Scot Class' locomotives were built between 1927 and 1930 at the North British Locomotive Works in Glasgow (50) and the LMS Works at Crewe (20). These powerful locomotives (6P), designed by Sir Henry Fowler, were built to haul express passenger trains on the West Coast Main Line.

All seventy of these locomotives were later rebuilt to a design by William Stanier and re-classified as 7P locomotives by BR in 1951. Numbers 46100 *Royal Scot* and 46115 *Scots Guardsman* are the only two 'Royal Scot Class' locomotives in preservation; the remainder were scrapped.

No. 46113 *Cameronian.* LMS 4-6-0 Fowler/Stanier 'Royal Scot Class' locomotive. Built in September 1927 at the North British Locomotive Company in Glasgow as LMS number 6113 (later, BR number 46113). Withdrawn from service in December 1962 whilst based at Leeds Holbeck Shed and scrapped at Crewe Works in June 1963. This locomotive was rebuilt by BR in December 1950. All other members of the class were also rebuilt. (Photo P.59)

No. 46122 *Royal Ulster Rifleman.* LMS 4-6-0 Fowler/Stanier 'Royal Scot Class' locomotive. Built in November 1927 at the North British Locomotive Company in Glasgow as LMS number 6122 (later BR 46122). Withdrawn from service whilst based at Carlisle Upperby Shed in October 1964. Scrapped by Drapers of Hull in February 1965. (Photo P.28)

No. 46137 *The Prince of Wales's Volunteers (South Lancashire).* LMS 4-6-0 Fowler/Stanier 'Royal Scot Class' locomotive. Built in October 1927 at the North British Locomotive Company in Glasgow as LMS number 6137 (later BR 46137). Withdrawn from service whilst based at Carlisle Upperby Shed in November 1962. Scrapped at Crewe Works in May 1963. (Photo P.51)

No. 46141 *The North Staffordshire Regiment.* LMS 4-6-0 Fowler/Stanier 'Royal Scot Class' locomotive. Built in October 1927 at the North British Locomotive Company in Glasgow as LMS number 6141 (later BR 46141). Withdrawn from service in April 1964 whilst based at Carlisle Upperby Shed and scrapped at Crewe Works in July 1964. This locomotive was rebuilt by BR in October 1950. (Photo P.103)

No. 46142 *The York and Lancashire Regiment.* LMS 4-6-0 Fowler/Stanier 'Royal Scot Class' locomotive. Built in October 1927 at the North British Locomotive Company in Glasgow as LMS number 6142 (later BR 46142). Withdrawn from service in January 1964 whilst based at Longsight Shed in Manchester and scrapped at Crewe Works later that month. This locomotive was rebuilt by BR in February 1951. (Photo P.51)

LMS 4-6-2 'Princess Royal Class' Stanier locomotives: general information

Just twelve 'Princess Royal Class' locomotives, designed by William Stanier, were built for the LMS Railway. The first two members of the class were introduced in 1933 and a further ten were built in 1935. These extremely powerful locomotives (power classification 8P) were specifically designed to work express passenger train services on the West Coast Main Line. Two of these locomotives, number 46201 (LMS 6201) *Princess Elizabeth* and number 46203 (LMS 6203) *Princess Margaret Rose* have been preserved.

No. 46208 *Princess Helena Victoria.* LMS 4-6-2 'Princess Royal Class' locomotive. Built at Crewe Works in August 1935 as LMS number 6208 (later BR 46208). Withdrawn from service in October 1962 whilst based at Liverpool Edge Hill Shed, and scrapped at Crewe Works in November 1962. (Photo P.106)

LMS 4-6-2 'Coronation Class' Stanier locomotives: general information

A total of thirty-eight 'Coronation Class' locomotives, designed by William Stanier, were built between 1937 and 1948. They were also referred to as 'Princess Coronation Class' and 'Duchess Class' locomotives.

These locomotives were a larger and improved version of the 'Princess Royal Class' locomotives and were the most powerful steam passenger locomotive ever built for the LMS Railway Company (Power classification LMS 7P or BR 8P).

They were specifically designed to haul new express passenger trains on the West Coast Main Line between London Euston and Glasgow Central. The most prestigious of these trains was named the '*Coronation Scot*' and was introduced in 1937 to commemorate the coronation of King George VI. It ran non-stop from London Euston to Glasgow Central, a distance of over 400 miles (640km), in 6 hours 30 minutes.

The first five 'Coronation Class' locomotives, numbers 6220–6224, were painted in Caledonian Railway Blue livery with silver horizontal speed lines, picked out in darker blue. The letters 'LMS' and the engine number were in block letters, also painted silver. These locomotives were streamlined with air-smoothed casing and were purpose built to operate the new '*Coronation Scot*' express passenger service. The passenger coaches were also painted Caledonian Railway Blue with silver horizontal lines to match the locomotives.

The other thirty-three 'Coronation Class' locomotives were painted in Crimson Lake Red (maroon) livery. Twenty-four of these were streamlined with air-smoothed casing and gold horizontal lines; the remainder were not streamlined.

Between 1946 and 1948, the air-smoothed casing was removed from all streamlined locomotives to facilitate ease of maintenance, and the whole class were later fitted with double chimneys and smoke deflectors. During the 1950s, the entire class was painted in BR Brunswick Green livery. Just three 'Coronation Class' locomotives have been preserved.

No. 46220 *Coronation*. LMS 4-6-2 Stanier 'Coronation Class' locomotive. Built at Crewe in June 1937 as LMS number 6220. Re-numbered 46220 by BR in 1948. Withdrawn from service in April 1963 and scrapped in May 1963. (Photo P.60)

No. 46222 *Queen Mary*. LMS 4-6-2 Stanier 'Coronation Class' locomotive. Built at Crewe in June 1937 as LMS number 6222 (later BR 46222). Withdrawn from service in October 1963 whilst based at Polmadie Shed, Glasgow, and scrapped at Crewe Works in November 1963. (Photo P.58)

No. 46229 *Duchess of Hamilton*. LMS 4-6-2 Stanier 'Coronation Class' locomotive. Built at Crewe Works in September 1938 as LMS number 6229 (later BR 46229). Withdrawn from service in February 1964 whilst based at Edge Hill Shed, Liverpool, and subsequently preserved as part of the National Collection by the National Railway Museum in York. (Photo P.44)

No. 46242 *City of Glasgow*. LMS 4-6-2 Stanier 'Coronation Class' locomotive. Built for BR at Crewe Works in May 1948 as BR number 46242. Withdrawn from service in October 1963 whilst based at Polmadie Shed in Glasgow, and scrapped at Crewe Works in November 1963. (Photo P.44)

LMS 2-6-0 Ivatt 'Class 2F' locomotive: general information

A total of 128 'Class 2F' 2-6-0 locomotives, designed by H. G. Ivatt, were built at Crewe, Darlington and Swindon between 1946 and 1953. They were low-powered tender locomotives, based on the earlier Ivatt 'Class 2' 2-6-2 tank locomotive. These locomotives were designed to work light goods and freight services and given an LMS power classification of 2F. Later however, they were used for working light passenger trains and were re-classified by BR as mixed traffic locomotives with a power classification of 2MT.

Due to their small size, they earned themselves the nickname 'Mickey Mouse' and formed the basis of the design for the BR Standard 'Class 2' tender locomotives which were designed by R. A. Riddles. A total of sixty-five of these standard class locomotives (power classification 2MT) were built at Darlington and introduced between December 1952 and 1956. They also had the nickname 'Mickey Mouse'.

No. 46413. LMS 2-6-0 Ivatt 'Class 2F' locomotive. Built at Crewe Works in

February 1947 as LMS number 6413 (later BR 46413). Re-classified by BR as a mixed traffic locomotive (2MT). Withdrawn from service in October 1965 whilst based at Ayr Shed in Scotland. Scrapped in February 1966 by G. H. Campbell, Atlas Works, Airdrie, Scotland. (Photo P.53)

No. 46453. LMS 2-6-0 Ivatt 'Class 2F' locomotive. Built at Crewe Works for BR in April 1950. Withdrawn from service April 1962 whilst based at Leeds Holbeck Shed. Scrapped at Darlington Works in September 1962. (Photo P.54)

No. 46481. LMS 2-6-0 Ivatt 'Class 2F' locomotive. Built at Darlington Works for BR in October 1951. Withdrawn from service in December 1962 whilst based at Malton Shed in North Yorkshire and scrapped at Crewe Works in May 1963. (Photo P.102)

LMS 0-6-0 Fowler 'Class 3F' 'Jinty' tank locomotive: general information

A total of 422, 0-6-0, 'Class 3F' tank locomotives were built for the LMS Railway between 1924 and 1931. They were commonly referred to as 'Jinty' locomotives or 'Jintys'.

These locomotives, designed by Henry Fowler, were based on his rebuilds of the former Midland Railway 'Class 2441' locomotives which were designed by Samuel Waite Johnson and introduced into service between 1899 and 1902. A total of sixty of these 'Class 2441' engines were built for the Midland Railway Company.

The 'Jinty' locomotives were extremely popular, and whilst designed primarily for working light goods trains, they were also ideal for shunting duties. At the outbreak of the Second World War, the Fowler tank engines were initially selected for use by the War Department as their standard shunting locomotives. A total of eight were also shipped to France for use in the war effort by the state-owned French Railways, but this practice ceased after the fall of France in 1940. Five of the eight returned to Britain after the war.

After nationalisation of the railway in 1948, BR found that these locomotives were ideal for light passenger duties as well as their use as light freight and shunting locomotives. Consequently, a number were fitted with push and pull apparatus and were frequently seen working rural and branch line passenger trains.

Of the 422 'Jinty' locomotives built, 417 survived in service until 1959 when their withdrawals started, although almost 200 were still being used in 1964. The last six members of the class were eventually withdrawn from service in 1967. Due to their large numbers and late survival, a total of nine of these engines were subsequently preserved.

No. 47406. LMS 0-6-0 Fowler 'Class 3F' 'Jinty' tank locomotive. Built by the Vulcan Foundry, Newton-le-Willows, Lancashire in December 1926 as LMS number 16489 (later LMS 7406). After working briefly at Warrington and Crewe, she moved to Carnforth in October 1928 where she worked for over thirty years. Allocated BR number 47406 in 1948 but did not display her new number until May 1950. Withdrawn from service for scrap in December 1966 and transported to Woodham Brothers Scrapyard in Barry, South Wales, where she languished for many years, until in 1983 she was bought by the Rowsley Locomotive Trust for restoration. She was painstakingly restored to her former glory and was returned to steam in 2010. Number 47406 is now fully operational and can be seen working as a preserved locomotive on the Great Central Heritage Railway in Loughborough. She is currently privately owned by Mr Roger Hibbert. (Photo P.77)

LMS (LNWR) 0-8-2T Bowen-Cooke 'Class 6F' tank locomotive: general information

A total of thirty 'Class 1185' 0-8-2 tank locomotives were built for the LNWR at Crewe Works between 1911 and 1917. They were designed by Charles Bowen-Cooke and were, in effect, a tank version of his 'G Class' tender locomotive introduced in

1906. All the 'Class 1185' tank engines were absorbed into the LMS Railway Company in 1923 and re-classified as LMS 'Class 6F' tank locomotives.

These locomotives were designed primarily for heavy shunting duties but were fitted with three-link couplings for goods work. In addition, a vacuum brake was installed to enable them to be used for passenger train workings.

Withdrawals of these locomotives started in 1934 and a total of nine survived to see the railways nationalised in 1948. Their life with BR however was a short one and the last member of the class was withdrawn from service in March 1953. The whole class were scrapped.

No. 47892. LMS (LNWR) 'Class 6F' 0-8-2T locomotive. Built at Crewe in January 1917 as LNWR 'Class 1185' number 714. Absorbed into LMS stock in 1923 as LMS 'Class 6F' number 7892 (later BR 47892). Withdrawn from service in February 1948 whilst based at Patricroft Shed, Manchester and scrapped at Crewe Works in June 1948. Although allocated a BR number in 1948, the number was never displayed as she was due to be withdrawn from service for scrap. (Photo P.33)

LMS Fowler 2-6-6-2T 'Beyer-Garratt Class' locomotives: general information

A total of thirty-three 'Garratt' type locomotives, constructed by the Beyer, Peacock Company and introduced by Henry Fowler, were built for the LMS between 1927 and 1930 for heavy freight duties. The first three were introduced in 1927 and the remaining thirty were introduced in 1930.

A 'Garratt' locomotive (commonly referred to as a 'Beyer-Garratt' due to being manufactured in the main by the Beyer, Peacock Company based on a design by Herbert Garratt) was articulated into three sections. The centre frame supported the locomotive boiler. A steam engine was mounted on a separate frame in front of the boiler and a second steam engine was mounted on another frame behind the boiler. The three articulated sections allowed the locomotive to negotiate sharp curves which would not be possible with a rigid locomotive.

The Garratt articulated design was first developed by the British locomotive engineer Herbert William Garratt in the early 1890s. He approached the firm of James Kitson of Hunslet, Leeds regarding production of the locomotive but his idea was rejected. However, Kitson's later accepted a similar design from Robert Stirling, for the building of articulated locomotives based on the design previously submitted by Herbert Garratt. This resulted in six such locomotives being built by the Kitson Company. The first was exported to Chile in 1894. A further two were exported to Rhodesia (modern day Zimbabwe) in 1903, followed by three to Jamaica in 1904. The original Kitson design was subsequently modified but in total, some fifty articulated locomotives were built by the company for export, the last being in 1935.

Following rejection by the Kitson Company, Herbert Garratt approached the Beyer, Peacock Company who did show an interest in his design although they insisted that some modifications be made by their own design team. Garratt agreed. Later, the Beyer, Peacock Company started building 'Beyer-Garratt Class' locomotives which would eventually see large scale production on a worldwide scale. In total, over 1,000 'Beyer-Garratt' locomotives were built by the Beyer, Peacock Company alone, the last order being from South Africa in 1968 as the firm was in the process of closing down.

The last known Beyer-Garratt type commercial locomotive to have been built was a narrow-gauge version built at the Phil Girdlestone Rail Company Workshops in Port Shepstone, South Africa. That locomotive was shipped to Argentina where it entered service in October 2006. A small number of Beyer-Garratt steam locomotives are still thought to be in operation worldwide, a lasting tribute to British steam locomotive engineering.

Other articulated steam locomotives which were similar in design to the 'Beyer-Garratt' locomotives were the 'Mallet' design locomotives invented by the Swiss engineer Anatole Mallet and used in many parts of the world. They ranged from narrow gauge locomotives favoured on Swiss Mountain Railways to the massive 2-10-10-2 locomotives used on the vast railways of the USA.

A lesser-known articulated locomotive was the 'Meyer' 0-4-0 + 0-4-0 locomotive, designed by Frenchman Jean-Jacques Meyer who patented the design in 1861. These locomotives were not as successful as the 'Garratt' or 'Mallet' locomotives and they were mainly confined to Europe. They were most commonly used in France, Germany and Switzerland.

Some aspects of the Meyer design were however incorporated into the design of the Robert Stirling locomotives (mentioned earlier) being produced by the Kitson Company in Leeds. After the change of design, the Kitson locomotives were re-named 'Kitson-Meyer' locomotives. The new 'Kitson-Meyer' locomotives were a successful design and turned out to be very popular on the Columbian and Chilean Railways in South America.

The Beyer-Garratt locomotives built for the LMS Railway Company and introduced by Henry Fowler had a wheel configuration of 2-6-6-2T (sometimes referred to as 2-6-0 + 0-6-2). They were not given a power classification by the LMS or by BR and as such, remained 'unclassified'. These locomotives were not without their problems. They were extremely heavy on coal consumption and quite frequently encountered axle bearing problems, in part due to their size and weight. All thirty-three members of the class were withdrawn from service between 1955 and 1958. They were all scrapped and unfortunately none was preserved.

No. 47977. LMS 2-6-6-2T 'Beyer-Garratt Class' locomotive. Built in September 1930 by Beyer, Peacock Ltd, Manchester as LMS number 7977 (later BR 47977). Withdrawn from service in June 1956 whilst based at Hasland Shed, Chesterfield, Derbyshire and scrapped at Crewe Works in July 1956. Originally built with a fixed coal bunker but later rebuilt with a revolving coal bunker and some other minor modifications. (Photo P.23)

No. 47981. LMS 2-6-6-2T 'Beyer-Garratt Class' locomotive. Built in October 1930 by Beyer, Peacock Ltd, Gorton, Manchester. Withdrawn from service in November 1956 whilst based at Toton Shed, Nottinghamshire. Scrapped at Crewe Works in January 1957. (Photo P.107)

No. 47983. LMS 2-6-6-2T 'Beyer-Garratt Class' locomotive. Built in October 1930 by Beyer, Peacock Ltd, Gorton, Manchester. Withdrawn from service in January 1956 whilst based at Hasland Shed, Chesterfield. Scrapped at Crewe Works in March 1956. (Photo P.23)

No. 47997. LMS 2-6-6-2T 'Beyer-Garratt Class' locomotive. Built in April 1927 by Beyer, Peacock Ltd, Gorton, Manchester. Withdrawn from service in February 1956 whilst based at Hasland Shed, Chesterfield. Scrapped at Crewe Works in May 1956. (Photo P.107)

LMS 2-8-0 Stanier 'Class 8F' locomotives: general information

A total of 852 'Class 8F' heavy freight locomotives, designed by William Stanier, were built between 1935 and 1946 and turned out to be extremely efficient and successful locomotives. They were, in effect, a more powerful freight version of his highly successful 'Black Five' locomotive.

At the outbreak of the Second World War, these locomotives were selected to become the standard design for wartime British freight locomotives. As a result, in addition to a total of 331 which were built for the LMS Railway Company, considerable numbers were built for the LNER, GWR and SR.

Large numbers were also commissioned by the War Department (WD) for use during the conflict. Many of these 'Class 8F' WD locomotives were shipped to the Middle East after the fall of France, to support the British Army which was engaged in hostilities in North Africa and eventually Italy. A number were also supplied to the Turkish Government for use on their state railways. In 1943 however, due to wartime economy sanctions being imposed, the 'Class 8F' building programme was halted and 2-8-0 'WD Austerity' locomotives were introduced as a cheaper alternative.

Not all the 'Class 8F' locomotives shipped abroad for the war effort arrived at their destinations as some were lost at sea during the sinking of vessels that were transporting them. A considerable number never returned to Britain after the war for various reasons. Some were destroyed during the conflict, others were sold to Middle Eastern Countries after the war had ended, whilst many were just scrapped or used for spare parts as it was not considered economically viable to transport them back to the UK.

After the railways were nationalised in 1948, a total of 666 'Class 8F' locomotives (including ex-WD locomotives) became part of the BR locomotive stock. They continued in their role as highly successful freight locomotives, all surviving into the 1960s. A total of 638 were still in service in 1965 and the last 150 were withdrawn during the last year of steam in 1968.

No. 48201. LMS 2-8-0 Stanier 'Class 8F' locomotive. Built in June 1942 by the North British Locomotive Company in Glasgow as LMS number 8201 (later BR 48201). Withdrawn from service in March 1968 and scrapped in July 1968. (Photo P.62)

No. 48202. LMS 2-8-0 Stanier 'Class 8F' locomotive. Built in July 1942 by the North British Locomotive Company in Glasgow as LMS number 8202. Withdrawn from service in July 1967 and scrapped in November 1967. (Photo P.21)

No. 48537. LMS 2-8-0 Stanier 'Class 8F' locomotive. Built in July 1945 at Doncaster Works as LMS number 8537. Withdrawn from service in October 1967 and scrapped in May 1968. (Photo P.21)

No. 48540. LMS 2-8-0 Stanier 'Class 8F' locomotive. Built in December 1944 at Darlington Works (LNER built) as LMS number 8540. Withdrawn from service in October 1967 and scrapped in May 1968. (Photo P.106)

No. 48622. LMS 2-8-0 Stanier 'Class 8F' locomotive. Built at Ashford in November 1943 as LMS number 8622. Withdrawn from service in October 1967 and scrapped in April 1968. (Photo P.63)

No. 48641. LMS 2-8-0 Stanier 'Class 8F' locomotive. Built for BR at Brighton in September 1949. Withdrawn from service in December 1966 and scrapped in February 1967. (Photo P.62)

LMS 'Class G2A' and associated 0-8-0 locomotives: general information

The LNWR built a number of different classes of 0-8-0 heavy freight locomotives from 1892 onwards. The prototype 0-8-0 heavy freight locomotive (number 2524) was a simple engine design by Francis William Webb which was introduced in 1892. A variety of 0-8-0 heavy freight locomotives followed as 'Classes' A, B, C, D, E, and F. These locomotives ranged from simple 2-cylinder engines to 3-cylinder compound engines and 4-cylinder compounds. The majority of locomotives in the classes mentioned above were not brand new locomotives, but instead were rebuilds and further rebuilds of existing locomotives. All the locomotives from the six classes mentioned above were eventually given further rebuilds or scrapped.

'Class G'. Locomotives from this class were 2-cylinder simple engines designed by George Whale and Francis Bowen Cooke. Thirty-two 'Class G' locomotives were rebuilds of 'Class B' (4-cylinder compounds) carried out between 1906 and 1917. In addition, sixty new 'Class G' locomotives were built between 1910 and 1912. 'Class G' locomotives were fitted with 160psi non-superheated boilers. All 'Class G' locomotives were later rebuilt as 'Class G1' locomotives, the first in 1912 (see 'Class G1' below) and the remaining ninety-one between 1920 and 1937, after which the 'Class G' was extinct.

'Class G1'. In 1912, 'Class G' locomotive 2653 (BR number 49154) was rebuilt with a superheated version of its 160psi boiler and became the first 'Class G1' locomotive to be introduced. 170 new 'Class G1' locomotives were then built between 1912 and 1918. In addition, 278 of the earlier engines were rebuilt with superheated boilers between 1912 and 1934 and they also became 'Class G1' locomotives, taking the grand total of 'Class G1' to 448.

'Class G2'. A total of sixty brand new 'Class G2' locomotives, designed by Hewitt Pearson Beames, were introduced between 1921 and 1922. They were similar to the 'Class G1' locomotive but had an even larger and more powerful 175psi superheated boiler which increased the power classification of the locomotive from 6F to 7F.

'Class G2A'. A further 327 'Class G1' locomotives were rebuilt into 'Class G2' locomotives after 1936 but instead of being re-classified as 'Class G2', they were given the classification 'Class G2A' in order to distinguish them from the sixty newly built 'Class G2' locomotives introduced in 1921/1922. Both the 'Class G2' and 'Class G2A' locomotives were given the nickname 'Duck Eight' locomotives.

No. 48921. LMS (LNWR) 8-8-0 'Class G2A 'locomotive. Built by the LNWR at Crewe Works in April 1902 as a 'Class G1' locomotive number 2567. Absorbed into The LMS in 1923 as LMS number 8921. Rebuilt into a 'Class G2A' locomotive in March 1940. Withdrawn from service in April 1958 and scrapped later that month. (Photo P.78)

No. 48930. LMS (LNWR) 0-8-0 'Class G2A' locomotive. Built by the LNWR at Crewe Works in April 1903 as LNWR 'Class B' locomotive number 1248, designed by Francis Webb. Re-numbered LMS 8930 in October 1928 (later BR 48930). Rebuilt as a 'Class G1' locomotive by Charles Bowen Cook in March 1932 and further rebuilt as a 'Class G2A' in March 1943. Withdrawn from service in December 1962 whilst based at Bescot Shed, Walsall, West Midlands and scrapped at Crewe Works in January 1963. (Photo P.90)

No. 49089. LMS (LNWR) 0-8-0 'Class G1' (G2A) locomotive. Built by the LNWR at Crewe Works in February 1910 as LNWR 'Class G' locomotive number 2662, designed by George Whale. Rebuilt as a 'Class G1' locomotive in October 1927 by Francis Bowen-Cooke and re-numbered LMS 9089 (later BR 49089). Further rebuilt as a 'Class G2A' in September 1938 but reverted back to a 'Class G1' locomotive in June 1940 for the remainder of her service. Withdrawn from service in April 1950 whilst based at Aston Shed, Birmingham but remained 'mothballed' until scrapped at Crewe Works in 1954. (Photo P.78)

LMS 0-8-0 'Class 7F' locomotives: general information

A total of 175, 0-8-0, 'Class 7F' locomotives, designed by Henry Fowler, were built between 1929 and 1932. They were nicknamed 'Baby Austins' or 'Austin 7s' after the Austin motor car which was popular at the time.

These 'Class 7F' locomotives were a development of the LNWR 'Class G2' locomotives. They were both economical and powerful but were, unfortunately, prone to problems with their wheel bearings and associated lubrication which resulted in 'hot axle boxes'. This frequently led to locomotives being taken out of service. Although the whole class entered BR service in 1948, their days were numbered, with over half the class being withdrawn for scrap by 1951. Their numbers were down to single figures by 1960 with the last one (number 49508) being withdrawn from service in January 1962. She was later scrapped at Crewe Works. The first member of the class (number 49395) has been preserved and is on public display at Shildon Railway Museum. All the rest were scrapped.

No. 49618. LMS 0-8-0 'Class 7F' locomotive. Designed by Henry Fowler and built at Crewe Works in April 1931 as LMS number 9618 (later BR 49618). Withdrawn from service in October 1961 whilst based at Mirfield Shed in West Yorkshire. Scrapped by the Central Wagon Company, Wigan in November 1962. (Photo P.89)

No. 49648. LMS 0-8-0 'Class 7F' locomotive. Designed by Henry Fowler and built at Crewe Works in February 1932 as LMS number 9648. Withdrawn from service in September 1957 whilst based at Agecroft Shed, Salford, Manchester. Scrapped at Horwich Works, Bolton in November 1957. (Photo P.90)

LMS 'Class 2P' and 'Class 3P' tank engines (LYR 'Classes 5 & 6'): general information

A total of 310, 2-4-2 tank locomotives, designed by John Aspinall, were built for the Lancashire and Yorkshire Railway Company between 1889 and 1911. They were designed for use as light passenger locomotives (power classification 2P). These locomotives were built in various batches, some of which included modifications. Initially they had a coal carrying capacity of two tons but this was increased to four tons from 1898 by the fitting of extended coal bunkers.

In 1905, George Hughes introduced Belpaire boilers and extended smokeboxes into the locomotives, and from 1910 he rebuilt some of the older engines with Belpaire boilers.

George Hughes continued to improve the 'Class 5' locomotives and between 1911 and 1914 he introduced twenty, new and more powerful, superheated versions of the locomotive which were classified as LYR 'Class 6' locomotives. Forty-four of the existing 'Class 5' locomotives were also rebuilt into 'Class 6' locomotives, taking the total to sixty-four. The 'Class 6' locomotives were given an increased power classification of 3P.

The Lancashire and Yorkshire Railway amalgamated into the London and North Western Railway on 1 January 1922. The following year they were absorbed into the newly formed LMS Railway. The former LYR 'Class 5' locomotives were re-classified as LMS 'Class 2P' locomotives. The former LYR 'Class 6' locomotives were re-classified as LMS 'Class 3P'.

A total of 278 LYR 'Class 5' locomotives were absorbed into the LMS in 1923 and 110 entered BR service in 1948. They continued working until 1961 when the last one was withdrawn from service. Just one example (LYR number 1008) has been preserved.

Withdrawals of LYR 'Class 6' locomotives started in 1928 and just fourteen survived until 1948 when they came under BR ownership. Half were withdrawn from service within the first twelve months of BR service and the last member of the class was withdrawn in 1952.

No. 50689. LMS 2-4-2T 'Class 2P' (LYR 'Class 5') locomotive. Designed by John Aspinall and built at Horwich Works in September 1893 as LYR locomotive number 119. Absorbed into LMS stock as LMS number 10689 (later BR 50689). Withdrawn from service in August 1952 whilst based at Manningham Shed, Bradford, West Yorkshire. Scrapped at Horwich Works shortly afterwards. (Photo P.36)

No. 50909. LMS 2-4-2T 'Class 3P' (LYR 'Class 6') locomotive. Designed by George Hughes and built at Horwich Works in May 1911 as LYR number 224. Later absorbed into LMS stock as LMS number 10909 (later BR 50909). Withdrawn from service in February 1951 whilst based at

Sowerby Bridge Shed, West Yorkshire and scrapped at Horwich Works in March 1951. (Photo P.36)

LMS 0-6-0 'Class 3F' (LYR 'Class 27'): general information

A total of 484 LYR 'Class 27' (also known as 'Class F19') 0-6-0, 2-cylinder standard goods locomotives, designed by John Aspinall, were built at Horwich Works from 1889 onwards for use on the Lancashire and Yorkshire Railway. Numerous variations and modifications were made to the class over a number of years as well as rebuilds which were also carried out. Following the promotion of John Aspinall to General Manager of the LYR in 1899, the 'Class 27' locomotive building programme continued under the auspices of his successors, Henry Hoy and George Hughes, who carried on with the building programme until 1918 when production ceased.

A total of 300 'Class 27' locomotives were absorbed into the LMS Railway in 1923 and they were later re-classified as LMS 'Class 3F' locomotives. Their popularity continued throughout the LMS era and as a tribute to their success, a total of 245 survived until 1948 when they entered BR service after the railways were nationalised. A considerable number remained operational throughout the 1950s until the last sixteen members of the class were withdrawn from service in 1961. One member of the class, number 52322 (LYR number 1300), was preserved by the Ribble Steam Railway and is currently based on the East Lancashire Heritage Railway. All other members of the class were scrapped.

No. 52154. LMS 0-6-0 'Class 3F' (LYR 'Class 27') locomotive. Designed by John Aspinall and built at Horwich Works in February 1892 as LYR number 1141. Absorbed into LMS stock as LMS number 12154 (later BR 52154). Withdrawn from service in November 1960 whilst based at Goole Shed in East Yorkshire. Scrapped at Horwich Works in January 1961. (Photo P.93 & 97)

No. 52305. LMS 0-6-0 'Class 3F' (LYR 'Class 27') locomotive. Designed by John Aspinall and built at Horwich Works in May 1897 as LYR number 449. Absorbed into LMS stock as LMS number 12305 (later BR 52305). Withdrawn from service in November 1960 whilst based at Goole Shed in East Yorkshire. Scrapped at Horwich Works in January 1961. (Photo P.94)

No. 52355. LMS 0-6-0 'Class 3F' (LYR 'Class 27') locomotive. Designed by John Aspinall and built at Horwich Works in December 1895 as LYR number 417. Absorbed into LMS stock as LMS number 12319 (later BR 52319). Withdrawn from service whilst based at Goole Shed in November 1960. (Photo P.45)

No. 52411. LMS 0-6-0 'Class 3F' (LYR 'Class 27') locomotive. Designed by John Aspinall and built at Horwich Works in April 1900 as LYR number 702. Absorbed into LMS stock as LMS number 12411 (later BR 52411). Withdrawn from service in November 1960 whilst based at Agecroft Shed in Salford, Greater Manchester. Scrapped at Crewe Works in January 1961. (Photo P.45)

No. 52435. LMS (LYR) 0-6-0 'Class 3F' locomotive. Designed by John Aspinall and built at Horwich Works in May 1901 as LYR 'Class 27' number 727. Absorbed into LMS locomotive stock in 1923 as LMS number 12435 (later BR 52435). Withdrawn from service in July 1954 and scrapped at Horwich Works. (Photo P.79)

LMS 0-8-0 'Class 7F' (LYR 'Class 31') locomotives: general information

A total of 115 'Class 31' 7F locomotives were built for the Lancashire and Yorkshire Railway between 1912 and 1920. In addition, forty 'Class 31' engines were rebuilt from existing locomotives, making a grand total of 155. These locomotives, designed by George Hughes, were a superheated version of the LYR 'Class 30' which was a '6F' locomotive, designed by John Aspinall and introduced between 1900 and 1918.

The 'Class 31' locomotives were absorbed into LMS Stock in 1923 and re-classified as LMS 'Class 7F' locomotives. Withdrawals of these locomotives started in 1926 and only seventeen survived into the BR era. The last three members of the class were withdrawn from service in 1952. All were scrapped and none was preserved.

No. 52857. LMS 0-8-0 'Class 7F' (LYR 'Class 31') locomotive. Designed by George Hughes and built at Horwich Works in June 1913 as LYR number 1563. Absorbed into LMS stock as LMS number 12857 (later BR number 52857). Withdrawn from service in December 1951 and scrapped at Horwich Works in May 1952. (Photo P.97)

LMS (HR) 4-6-0 'Class 4P' ('Clan Class') locomotives: general information

Just eight 4-6-0 'Clan Class' locomotives, designed by Christopher Cumming, were built for the Highland Railway Company. Four were built in 1919 and a further four in 1921. They were built by Hawthorne Leslie and Company of Newcastle.

These eight locomotives were designed by Cumming exclusively for use as express passenger locomotives, although they were similar in design to his 'Clan Goods Class' engines which had been introduced the previous year. The passenger locomotives were slightly larger and heavier than his goods engines and were fitted with 6ft 0in (1.829m) driving wheels as opposed to 5ft 3in (1.600m) driving wheels on his goods engines.

Initially, six 'River Class' locomotives, designed by the HR Chief Mechanical Engineer, Mr F. G. Smith (predecessor to Christopher Cumming) were built in 1915 to work express passenger trains on the Highland Railway. Whilst they were the most powerful locomotives ever built for the company, there was a dispute between Smith and the Chief Civil Engineer, Alexander Newlands, as to whether or not the locomotives exceeded the maximum weight allowed for locomotives to operate on the Highland Railway. The dispute culminated when Smith was dismissed from the company and replaced by Cumming. Smith's 'River Class' locomotives were never used on the Highland Railway but were sold to the Caledonian Railway Company where they proved themselves to be very successful until the last one was withdrawn from service in December 1946. The eight HR 'Clan Class' locomotives introduced by Cumming were withdrawn from service and scrapped between 1944 and 1950.

No. 54767 *Clan MacKinnon*. LMS (HR) 4-6-0 'Clan Class' locomotive. Built in July 1921 by Hawthorne Leslie and Company, Newcastle, as Highland Railway locomotive number 55 *Clan MacKinnon*. Absorbed into the LMS in 1923 and re-classified as LMS 'Class 4P' number 14767. Entered BR service in 1948 as BR number 54767. She was withdrawn from service in January 1950 and scrapped in February 1950. This locomotive was one of just two members of the class which survived into the BR era and the only one to display a BR number. The other locomotive, LMS number 14764 *Clan Munro*, was allocated BR number 54764 but was withdrawn from service for scrap in February 1948 before she actually received her number. (Photo P.59)

LMS (CR) 0-6-0 'Class 3F' locomotives: general information

A total of seventy-nine Caledonian Railway 'Class 812' locomotives were built in six batches between 1899 and 1900. The first batch (seventeen locomotives) were turned out in the famous Caledonian Blue livery. All remaining members of the class were painted black with red lining.

These locomotives, designed by John McIntosh, were in fact an enlarged and more powerful version of the CR 'Class 294' ('Jumbo Class') 2F locomotives designed by Dugald Drummond and introduced between 1883 and 1897.

Caledonian Railway 'Class 812' locomotives were designed primarily as freight locomotives (power classification 3F), to be used for working local goods trains. They were however quite versatile and were frequently used as mixed traffic locomotives. They worked suburban passenger trains in and out of Glasgow as well as working local passenger services elsewhere. It was not uncommon to see them engaged in shunting duties as well as working local freight which they were designed for.

Between 1908 and 1909, John McIntosh introduced seventeen new CR 'Class 652' locomotives. Apart from the design of the locomotive cab, these locomotives were identical to his 'Class 812' locomotives.

Of the ninety-six 'Class 812' and 'Class 652' locomotives built by John McIntosh, ninety-three survived to see BR service and the last withdrawals did not take place until 1963.

All ninety-six were scrapped, with the exception of BR number 57566 which was preserved and can be seen working on the Strathspey Heritage Railway in Scotland.

No. 57626. LMS (CR) 'Class 3F' 0-6-0 locomotive. Built in October 1899 at St Rollox Engineering Works, Glasgow as Caledonian Railway 'Class 812' locomotive number 291. Absorbed into the LMS in 1923 and re-classified as LMS 'Class 3F' number 7626. Entered BR in 1948 as BR 'Class 3F' number 57626. Withdrawn from service in February 1962 and scrapped in October 1962. (Photo P.58)

LMS unclassified locomotive number 58100: general information

In 1919, the largest locomotive ever built for the Midland Railway emerged from Derby Locomotive Works. It was carrying the Midland Railway number 2290 and although the locomotive was not officially named, it quickly became known as *Big Bertha*.

Big Bertha was designed by Henry Fowler and James Anderson and was very much a one-off design. She was purpose-built to be used as a banking engine, to assist trains in ascending the Lickey incline by pushing them from the rear. The banking procedure also, in the event of a coupling failure, prevented coaches or wagons breaking away from a train and rolling back down the incline into the path of other oncoming trains. The Lickey incline itself, was and still is, the steepest sustained stretch of main line in Britain. It is located at Bromsgrove in Worcestershire on the Birmingham to Bristol (former Midland Railway) main line. The incline stretches for over two miles (3.2km) with a gradient of 1 in 37.

Big Bertha was a massive 4-cylinder locomotive which weighed over 100 tons (100,000kg). Her appearance was very distinct, with her ten driving wheels and huge inclined piston cylinders. She was an extremely powerful locomotive built for one specific purpose and as such she was not suitable for working either freight or passenger trains. Consequently, she was never awarded a power classification. Attached to the top of the smokebox at the front of the locomotive was a huge electric headlight to assist *Big Bertha* in attaching herself to the rear of trains during the hours of darkness. Number 58100 *Big Bertha* was successfully used almost continually from 1919 to assist both goods and passenger trains up the Lickey incline until 1956, when she was withdrawn from service having travelled almost 840,000 miles. She was later scrapped. During her working life *Big Bertha* was considered to be almost indispensable and as such a spare boiler and other parts were retained at Derby Works in order to minimise the length of time she would be out of service for routine maintenance, servicing or overhaul.

No. 58100. LMS 0-10-0 Class 'un-classified' locomotive. Built at Derby Works in December 1919 as Midland Railway number 2290. Although the locomotive was absorbed into LMS stock in 1923, she retained her MR number until 1947 when she was allocated an LMS number

22290. The following year, the railways were nationalised, and she was given a BR number 58100 in January 1949. Withdrawn from service in May 1956 whilst based at Bromsgrove Shed. Scrapped at Derby Works in September 1957. Although never officially named, 58100 was commonly referred to as *Big Bertha*. (Photo P.94)

LMS (MR) 0-6-0 'Class 2F': general information

A total of 935, 0-6-0 'Class 2F' tender locomotives, designed by Samuel Johnson, were introduced as new or rebuilt locomotives for the Midland Railway between 1875 and 1902. This was numerically one of the largest classes of locomotives to serve on Britain's railways.

Such 0-6-0 locomotives were the standard design freight locomotives used by the Midland Railway Company. The locomotives were introduced over the twenty-seven-year period in numerous batches. Progressive batches usually contained minor modifications. Large numbers of these locomotives were successively rebuilt over the years and many, which were fitted with bigger boilers, were re-classified as 3F locomotives.

After the 1923 railway groupings, all these locomotives were absorbed into LMS stock, but withdrawals started as early as 1925 and continued for almost thirty years. When the railways were nationalised in 1948, the rebuilt 3F locomotives were renumbered in line with other LMS engines but the remaining 2F locomotives were given a new series of BR numbers, 58114 to 58310. Many of these locomotives survived the 1950s, until the last one was withdrawn for scrap in 1964. Despite the massive numbers built, not one of these engines was preserved.

No. 58198. LMS (MR) 0-6-0 'Class 2F' locomotive. Built at Derby Works in March 1880 as Midland Railway number 3047. Rebuilt after 1917 with a non-superheated Belpaire boiler and entered LMS Stock in 1923. Later became BR number 58198 until being withdrawn from service in September 1959 whilst based at Canklow Shed, Rotherham, South Yorkshire. She was scrapped at Derby Works in July 1960. (Photo P.19)

LMS 0-6-0 'Class 2F' (NLR 'Class 75'): general information

A total of thirty North London Railway 'Class 75' 0-6-0 tank locomotives were built at Bow Locomotive Works, East London between 1879 and 1905. They were designed by John Carter Park to be used as shunting locomotives in the East India and West India Docks in London which at the time were being serviced by the North London Railway Company.

Nine of the locomotives were transferred to the London and North Western Railway in 1909 and the remaining twenty-one later followed suit. All thirty locomotives were absorbed into the LMS in 1923 and re-classified as LMS 'Class 2F' locomotives. Withdrawals started in 1930 and just fourteen survived into the BR era. The last member of the class (number 58850) was withdrawn from service in 1960.

No. 58850. LMS (NLR) (LNWR) 0-6-0T 'Class 2F' tank locomotive. Built at the North London Railway Locomotive Works in 1880 as NLR 'Class 75' number 76. Re-numbered 116 in 1891 before being rebuilt at Bow in 1897. Went on loan to the LNWR in 1909 as number 2650 before being transferred into the stock of the newly formed LMS Railway in 1923. She was allocated an LMS number, 7505, in 1926 and under a re-numbering scheme, she became LMS number 27505 in June 1934. She received her BR number, 58850, in May 1949. From the early 1930s until her withdrawal from service in September 1960, 58850 was based at Rowsley Shed in Derbyshire, from where she performed freight work and shunting duties on the Cromford and High Peak railway line.

After being withdrawn from service, number 58850 was stored at Derby before being sold to the Bluebell Heritage Railway for preservation in 1962. She continues to be a part of the Bluebell Railway. (Photo P.50)

LMS 'Class 2F' (LNWR 'Webb coal tank') locomotives: general information

A total of 300 'Webb coal tanks' were built for the LNWR between 1881 and 1897. They were designed by William Francis Webb as a tank version (power classification 2F) of his earlier tender version 0-6-0, 'coal engine'. As the name suggests, they were primarily designed for hauling coal and other heavy mineral trains, particularly coal from the South Wales Coalfields and slate from the quarries of North Wales. They did however prove themselves to be quite versatile locomotives and were later used all over the LNWR network for working rural branch line and local passenger services. Webb used cast iron spoked wheels on these engines to reduce manufacturing costs.

Withdrawals of these locomotives commenced in 1921 and 292 were absorbed into LMS stock in 1923. Sixty-four survived into the BR era and lasted until 1958.

No. 58926. LMS (LNWR) 0-6-2T 'Class 2F' tank locomotive. Built at Crewe Works in September 1888 as LNWR number 1054. Absorbed into the LMS in 1923 as LMS number 7799 then 27799 (later BR number 58926).

Number 58926 was withdrawn from service in January 1939 and was destined for the scrapyard when lady luck intervened. After standing idle for a number of months awaiting her fate, 58926 had an unexpected 'eleventh hour reprieve'. This resulted from the outbreak of the Second World War. Due to the need for additional locomotives for the war effort, instead of being scrapped, 58926 was given a complete overhaul and put back into service.

She subsequently survived the war, entered BR service in 1948 and continued working for a further decade. 58926 was the last surviving member of her class when she was finally withdrawn from service for scrap in November 1958.

Amazingly, history repeated itself for a second time and 58926 had a further reprieve from the scrapyard after a public appeal was launched to raise the princely sum of £666 to purchase her for preservation. The appeal was successful and she is currently owned by the National Trust. Number 58926 can be seen working in all her glory at the Keighley and Worth Valley Heritage Railway in West Yorkshire. She is without doubt one very lucky locomotive. (Photo P.50)

LONDON AND NORTH EASTERN RAILWAY (LNER) LOCOMOTIVES

LNER locomotives pictured in this book displaying their old LNER numbers are set out below. Information about these engines is listed under their respective BR numbers.

No. 500 – See BR number 60500
No. 888 – See BR number 60888
No. 1179 – See BR number 61179
No. 2255 – See BR number 62255
No. 2737 – See BR number 62737
No. 4203 – See BR number 64203
No. 7356 – See BR number 67356
No. 8359 – See BR number 68359
No. 9446 – See BR number 69446
No. E9446 – See BR number 69446

The remaining locomotives pictured in this book are all displaying their BR numbers, under which all details and technical information about each locomotive can be found.

LNER 4-6-2 'Class A4' locomotives: general information

A total of thirty-five 'Class A4' Streamlined 'Pacific' locomotives, designed by Nigel Gresley, were built at Doncaster between 1935 and 1938. They were a development of his 'Class A3' locomotives but with a radical new appearance. The locomotives were covered in a streamlined casing with a wedge-shaped front and side valances. Designed and built as express passenger locomotives to work the East Coast Main Line between London (King's Cross) and Edinburgh via York and Newcastle, also King's Cross and Leeds, they were given the nickname 'Streaks' due to their streamlined appearance and fast speed.

The first four 'Class A4' locomotives built were each painted silver-grey and given names which contained the word silver. They were number 2509 *Silver Link*, 2510 *Quicksilver*, 2511 *Silver King* and 2512 *Silver Fox*. These four locomotives were selected to work a new daily express passenger train service on the East Coast Main Line between London King's Cross and Newcastle. The train service was introduced on 30 September 1935 to commemorate the silver jubilee of King George V which was being celebrated that year. The passenger train, which was named '*Silver Jubilee*', consisted of seven coaches (eight coaches from 1938), which were also painted silver-grey to match the silver-grey livery of the locomotives. The first journey of the new service was a complete success and the train, which was hauled by number 2509 *Silver Link*, attained a speed of 112mph (180km/h) and completed the journey in just four hours at an average speed of 67mph (108km/h).

'Class A4' locomotives quickly became synonymous with speed and a pinnacle was achieved when number 4468, *Mallard*, set a world record for the fastest steam locomotive when she attained a speed of 126mph (203km/h) on 3 July 1938. The record has never been surpassed.

A total of six 'Class A4' locomotives have been preserved as follows: 60007 *Sir Nigel Gresley*, 60008 *Dwight D. Eisenhower* (donated to the National Railroad Museum, USA), 60009 *Union of South Africa*, 60010 *Dominion of Canada* (donated to the Canadian National Railway Museum), 60019 *Bittern* and 60022 *Mallard*.

No. 60001 *Sir Ronald Matthews.* LNER 4-6-2 'Class A4' locomotive. Built at Doncaster in April 1938 as LNER number 4500. She was given the name *Garganey* (named after a wild duck). In March 1939 she was re-named *Sir Ronald Matthews*, who had been appointed Chairman of the LNER Board of Directors in October 1938. Under an LNER re-numbering scheme, her number was changed from 4500 to LNER number 1 in 1946, before being allocated BR number 60001 after the railways were nationalised in 1948. She was withdrawn from service in October 1964 whilst based at Gateshead, and scrapped by Hughes Bolckows of North Blyth in January 1965. (Photo P.52)

No. 60002 *Sir Murrough Wilson.* LNER 4-6-2 'Class A4' locomotive. Built at Doncaster in April 1938 as LNER number 4499 *Pochard* (named after a wild duck). In April 1939 she was re-named *Sir Murrough Wilson* after a former Director of the North Eastern Railway Company, former British Army officer in the West Yorkshire Regiment and former MP for the constituency of Richmond in North Yorkshire.

Under an LNER re-numbering scheme, number 4499 was changed to LNER number 2 in 1946, before being allocated BR number 60002 after the railways were nationalised in 1948. She was withdrawn from service in May 1964 whilst based at Gateshead and scrapped by Cohens of Middlesbrough in November 1964. (Photo P.13)

No. 60009 *Union of South Africa.* LNER 4-6-2 'Class A4' locomotive. Built at Doncaster in June 1936 as LNER number 4488 *Osprey* (named after a duck), which she carried until she was renamed *Union of South Africa* the following year. Her number was changed from LNER 4488 to LNER number 9 in 1946, followed by BR number 60009, which she carried from 1948 until she was withdrawn from service in June 1966. She was subsequently preserved and has since been used on the main-line network to haul special excursion trains for steam enthusiasts and members of the public. She is privately owned by Mr John Cameron, a Scottish farmer. (Photo P.15)

No. 60016 *Silver King.* LNER 4-6-2 'Class A4' locomotive. Built at Doncaster in November 1935 as LNER number 2511 *Silver King*. Under an LNER re-numbering scheme, her number was changed from 2511 to LNER number 16 in 1946 before being allocated BR number 60016 after the railways were nationalised in 1948. She was withdrawn from service in March 1965 whilst based at Aberdeen Ferryhill and scrapped by the Motherwell Machinery & Scrap Company, Scotland in May 1965. When introduced into service, number 60016 was one of four locomotives painted silver to operate the *'Silver Jubilee'* express passenger service which ran between London King's Cross and Newcastle. (Photo P.40)

No. 60031 *Golden Plover.* LNER 4-6-2 'Class A4' locomotive. Built at Doncaster in October 1937 as LNER number 4497. She was given the name *Golden Plover* after a species of wading bird. Under an LNER re-numbering scheme, her number was changed from 4497 to LNER number 31 in 1946 before being allocated BR number 60031 after the railways were nationalised in 1948. She was withdrawn from service in November 1965 whilst based at St Rollox Shed, Glasgow, and scrapped by G. H. Campbell of Airdrie in February 1966. (Photo P.52)

LNER 4-6-2 'Class A3' locomotives: general information

LNER 'Class A3' locomotives were a development of the GNR 'Class A1' locomotives designed by Nigel Gresley. A total of fifty-two (GNR design) 'Class A1' locomotives were built between 1922 and 1935. In 1923, the GNR was absorbed into the LNER and Nigel Gresley was appointed Chief Mechanical Engineer of the newly formed LNER Company.

Gresley quickly developed a new and more powerful locomotive, the LNER 'Class A3', to supersede his 'Class A1' locomotives. Fifty-one of his 'Class A1' locomotives were subsequently rebuilt as 'Class A3' locomotives between 1927 and 1948. In addition, a further twenty-seven brand new 'Class A3' locomotives were built during the same period, taking the grand total of 'Class A3' locomotives to seventy-eight.

The remaining 'Class A1' locomotive, number 4470 (BR 60113) *Great Northern* (the first 'Class A1' to be built), was not rebuilt as a 'Class A3' but was rebuilt by Edward Thompson in 1945, retaining the classification 'Class A1' (rebuild). Edward Thompson was the successor to Gresley, who died suddenly in 1941 after a short illness.

In April 1945, sixteen former GNR 'Class A1' locomotives which had not yet received their rebuilds into 'Class A3' locomotives were re-classified as LNER 'Class A10' locomotives and remained as such until their rebuilds to 'Class A3' were carried out. This was to make way for a new class of A1 locomotives which were being designed and developed by Arthur Peppercorn.

No. 60048 *Doncaster.* LNER 4-6-2 'Class A3' locomotive. Built at Doncaster in August 1924 as LNER 'Class A1' number 2547. She was given the name *Doncaster*, not after the town in South Yorkshire where she was built, but after a racehorse who won the 1873 Epsom Derby. She was rebuilt as a 'Class A3' locomotive in May 1946 and numbered LNER number 48. Allocated BR number 60048 after the railways were nationalised in 1948. Withdrawn from service in September 1963 whilst based at Grantham Shed and scrapped at Doncaster Works in October 1963. (Photo P.101)

No. 60055 *Woolwinder.* LNER 4-6-2 'Class A3' locomotive. Built at Doncaster in December 1924 as LNER 'Class A1' locomotive number 2554. She was given the name *Woolwinder* after the racehorse 'Wool Winder' (two words), who won the St Leger in 1907. Rebuilt as a 'Class A3' locomotive in June 1942 and re-numbered LNER number 55 in 1945. Allocated BR number 60055 after the railways were nationalised in 1948. Withdrawn from service in September 1961 whilst based at King's Cross Top Shed and cut up at Doncaster Works shortly afterwards. (Photo P.75)

No. 60063 *Isinglass.* LNER 4-6-2 'Class A3' locomotive. Built at Doncaster in June 1925 as LNER 'Class A1' locomotive number 2562. She was given the name *Isinglass* after a racehorse who won the Derby, the St Leger and the 2,000 Guineas in 1893. Rebuilt as an LNER 'Class A3' locomotive, number 63, in April 1946. Allocated BR number 60063 after the railways were nationalised in 1948. Withdrawn from service in June 1964 whilst based at New England Shed, Peterborough, and scrapped by A. King & Son, Norwich in October 1964. (Photo P.34)

No. 60067 *Ladas.* LNER 4-6-2 'Class A3' locomotive. Built at the North British Locomotive Works, Glasgow in August 1924 as LNER 'Class A1' locomotive number 2566. She was given the name *Ladas,* after a racehorse who won the Derby in 1894. Rebuilt as a 'Class A3' locomotive in November 1939 and re-numbered as LNER number 67 in 1945. Allocated BR number 60067 after the railways were nationalised in 1948. Withdrawn from service in December 1962 whilst based at King's Cross Top Shed, and scrapped at Doncaster Works in January 1963. (Photo P.75)

No. 60071 *Tranquil.* LNER 4-6-2 'Class A3' locomotive. Built at the North British Locomotive Works, Glasgow in September 1924 as LNER 'Class A1' locomotive number 2570. She was given the name *Tranquil,* after a racehorse who won the St Leger in 1923. Rebuilt as a 'Class A3' locomotive in October 1944 and re-numbered LNER number 71 in 1945. Allocated BR number 60071 after the railways were nationalised in 1948. Withdrawn from service in October 1964 whilst based at Gateshead and scrapped by Drapers of Hull in December 1964. (Photo P.40)

No. 60076 *Galopin.* LNER 4-6-2 'Class A3' locomotive. Built at the North British Locomotive Works, Glasgow in October 1924 as LNER 'Class A1' locomotive number 2575. She was given the name *Galopin,* after a racehorse who won the Derby in 1875. Rebuilt as a 'Class A3' locomotive in June 1941 and re-numbered as LNER number 76 in 1945. Allocated BR number 60076 after the railways were nationalised in 1948. Withdrawn from service in October 1962 whilst based at Heaton Shed, Newcastle and scrapped at Doncaster Works in April 1963. (Photo P.76)

No. 60078 *Night Hawk.* LNER 4-6-2 'Class A3' locomotive. Built at the North British Locomotive Works, Glasgow in October 1924 as LNER 'Class A1' locomotive number 2577. She was given the name *Night Hawk* after a racehorse who won the St Leger in 1913. Rebuilt as a 'Class A3' locomotive in January 1944 and re-numbered as LNER number 76 in 1945. Allocated BR number 60078 after the railways were nationalised in 1948. Withdrawn from service in October 1962 whilst based at Heaton Shed, Newcastle and scrapped at Doncaster Works in April 1963. (Photo P.101)

No. 60102 *Sir Frederick Banbury.* LNER 4-6-2 'Class A3' locomotive. Built at Doncaster in July 1922 as LNER 'Class A1' locomotive number 4471. She was given the name *Sir Frederick Banbury* after a prominent businessman and MP who was the last chairman of the Great Northern Railway before it was absorbed into the LNER in 1923. He was a strong opponent of the railway groupings.

Number 60102 was rebuilt as an LNER 'Class A3' locomotive in October 1942 and re- numbered as LNER number 63 in April 1946. Allocated BR number 60063 after the railways were nationalised in 1948. Withdrawn from service in November 1961 whilst based at King's Cross Top Shed and scrapped at Doncaster Works later that month. (Photo P.34)

No. 60107 *Royal Lancer.* LNER 4-6-2 'Class A3' locomotive. Built at Doncaster in May 1923 as LNER 'Class A1' locomotive number 4476. She was given the name *Royal Lancer* after a racehorse who won the St Leger in 1922. Rebuilt as a 'Class A3' locomotive in October 1946 and re-numbered as LNER number 107. Allocated BR number 60107 after the railways were nationalised in 1948. Withdrawn from service in September 1963 whilst based at Grantham Shed and scrapped in November 1963 at Doncaster Works. (Photo P.33)

No. 60108 *Gay Crusader.* LNER 4-6-2 'Class A3' locomotive. Built at Doncaster in June 1923 as LNER 'Class A1' locomotive number 4477. She was given the name *Gay Crusader* after a racehorse who won the 2,000 Guineas and the St Leger in 1917. Rebuilt as a 'Class A3' locomotive in January 1943 and re-numbered as LNER number 108 in 1945. Allocated BR number 60108 after the railways were nationalised in 1948. Withdrawn from service in October 1963 whilst based at Doncaster Shed and scrapped at Darlington Works in January 1964. (Photo P.102)

LNER 'Class A1' locomotive 'Peppercorn' design: general information

A total of forty-nine 'Class A1' locomotives, originally designed by Arthur Peppercorn for use on the LNER, were introduced onto British Railways after nationalisation in 1948. They were built at Doncaster and Darlington between 1948 and 1949. These locomotives should not be confused with the earlier Gresley LNER (GNR) 'Class A1' locomotives built between 1922 and 1935, as they were a completely different design.

The Peppercorn A1 locomotives were however based on a modified version of the Nigel Gresley 'Class A1' design. The modified version being locomotive number 60113, *Great Northern,* after it had been rebuilt by Edward Thompson in 1945.

The Peppercorn 'Class A1' locomotive was designed as a powerful passenger locomotive (power classification BR 8P) and built to haul what were the heaviest post war passenger trains in Britain at the time, on the East Coast Main Line between London and Aberdeen via York and Edinburgh. Originally, none of the locomotives had names but eventually nameplates were fitted to the smoke deflectors of the whole class. Towards the end of the steam era, a number of these 'Class A1' locomotives, along with other passenger steam engines, were often seen working freight services, a job they were never designed for. All the Peppercorn 'A1' locomotives were withdrawn from service between 1962 and 1966. None was preserved, although a brand new locomotive of this class, number 60163 *Tornado*, was built in 2008 and she can be seen operating steam enthusiasts' and special excursion trains on the main line network.

No. 60127 *Wilson Worsdell*. LNER Peppercorn design 4-6-2 'Class A1' locomotive. Built at Doncaster in May 1949 as BR number 60127. Withdrawn from service in June 1965 whilst based at Gateshead Shed and scrapped by Hughes Bolckows of North Blyth in November 1965. (Photo P.27)

No. 60130 *Kestrel*. LNER Design 4-6-2 'Class A1' locomotive. Built at Darlington in September 1948 as BR number 60130. Withdrawn from service in October 1965 whilst based at Ardsley Shed, West Yorkshire and scrapped later that month by Cashmores, Great Bridge, Tipton, West Midlands. (Photo P.76)

LNER 'Class A2/3' 'Pacific' locomotive: general information
A total of fifteen 'Class A2/3' locomotives, designed by Edward Thompson, were built at Doncaster during 1946/47. A further fifteen were due to be built but following the retirement of Thompson, they were re-designed by his successor Arthur Peppercorn and introduced into BR in 1947/48 as new Peppercorn, 'Class A2' mixed traffic locomotives (power classification 8P7F).

LNER 'Class A2' 'Pacific' locomotives – Sequence of production and designers.
LNER (NER) Raven new design **'Class A2'** locomotives. Five were built between 1922 and 1924 for the North Eastern Railway Company and absorbed into the LNER after 1923.

LNER Thompson design **'Class A2/1'** locomotives (Rebuilds of Gresley 'Class V2' mixed traffic locomotives). Four were built in 1944.

LNER Thompson design **'Class A2/2'** locomotives (Rebuilds of Gresley 'Class P2' locomotives). Six were built between 1943 and 1945.

LNER Thompson new design **'Class A2/3'** locomotives. Fifteen were built in 1946.

LNER/BR Peppercorn new design **'Class A2'** locomotives. Fifteen built for BR between 1947 and 1948.

No. 60500 *Edward Thompson*. LNER 4-6-2 'Class A2/3' locomotive. Built at Doncaster in May 1946 as LNER number 500 (BR number 60500 after 1948). Withdrawn from service in June 1963 whilst based at New England Shed, Peterborough, and scrapped at Doncaster Works in August 1963. This engine was named after the LNER engineer who designed her. (Photo P.108)

No. 60508 *Duke of Rothesay*. LNER 4-6-2 'Class A2/1' locomotive. Built at Darlington in June 1944 as LNER number 3697. Re-numbered LNER number 508 in 1945 and given BR number 60508 after nationalisation in 1948. Withdrawn from service in February 1961 whilst based at New England Shed, Peterborough, and scrapped at Doncaster Works in April 1961.

Number 60508 was involved in a serious accident. On 17 July 1948, a fireman working LNER locomotive number 60508 *Duke of Rothesay* was killed and the driver injured, when the locomotive derailed whilst hauling an eleven-coach express passenger train from Edinburgh to King's Cross through Barnet Tunnel in Hertfordshire on the East Coast Main Line. The locomotive detached itself from the train, turned onto its side and came to rest just north of New Southgate Station where the passenger coaches subsequently derailed on a set of points. Eleven passengers were also injured.

An enquiry later revealed that the accident was caused by track defects resulting from inadequate track maintenance during the war years. The train was travelling at 70mph when the accident occurred. The locomotive was later salvaged and returned to service after extensive repairs had been carried out. (Photo P.77)

No. 60516 *Hycilla*. LNER 4-6-2 'Class A2/3' locomotive. Built at Doncaster in November 1946 as LNER number 516 (later BR 60516). Withdrawn from service in October 1962 whilst based at York North Shed and scrapped at Doncaster Works in April 1963. (Photo P.108)

LNER 'Class V2' locomotives: general information

A total of 184 LNER 'Class V2' locomotives, designed by Nigel Gresley, were built at Darlington and Doncaster between 1936 and 1944. They were designed for use as mixed traffic locomotives to haul express freight trains and passenger services. They were given a power classification 6MT which was later re-classified as 7P6F. The first member of the class, number LNER 4771 (BR 60800), was given the name *Green Arrow*, after an express freight service which had been operating on the LNER since 1928. The service was discontinued after the outbreak of war in 1939 but reinstated by BR in 1953. She is the only member of the class to have been preserved. 'Class V2' locomotives were often referred to as the Green Arrow Class.

In total, just eight of the 184 'Class V2' locomotives built were given names. Apart from *Green Arrow* mentioned above, five were named after British Army regiments and the other two after public schools located in Durham and York (St Peter's), two cities closely associated with the LNER. The whole class was withdrawn between 1962 and 1966.

No. 60878. LNER 2-6-2 'Class V2' locomotive. Built at Doncaster in July 1940 as LNER number 4849. Re-numbered as LNER 878 in 1945 and given BR number 60878 after nationalisation in 1948. Withdrawn from service in October 1962 whilst working out of York North Shed and scrapped at Darlington BR Works in November 1962. (Photo P.39)

No. 60888. LNER 2-6-2 'Class V2' locomotive. Built at Darlington in December 1939 as LNER number 4859. Re-numbered as LNER 888 in 1945 and given BR number 60888 after nationalisation in 1948. Withdrawn from service in December 1962 whilst working out of Aberdeen Ferryhill Shed, and scrapped in September 1963 at Cowlairs Locomotive Works in Glasgow. (Photo P.99)

LNER 'Class B1' locomotives: general information

A total of 410 LNER 'Class B1' locomotives, designed by Edward Thompson, were built between 1942 and 1952. These 2-cylinder locomotives were designed as mixed traffic locomotives to be on a par with the GWR 'Hall Class' and the LMS 'Black Five' locomotives but were of lesser quality due to cost cutting and austerity measures imposed during and after the Second World War. Nevertheless, they were popular locomotives which performed well and proved themselves to be very reliable and efficient mixed traffic locomotives (5MT).

Of the 410 'Class B1' locomotives built, fifty-nine were given names. The first forty members of the class were named after breeds of the South African antelope, which gave rise to the whole class being nicknamed 'bongos' or 'antelopes'. A further eighteen were named after LNER company directors, many of whom were unfamiliar names to the public at large. One other 'Class B1' named locomotive was number 61379, *Mayflower*. She was built in 1951 and given the name *Mayflower* to commemorate the Pilgrim Fathers' voyage to America in 1620 on a sailing ship by that name. Just two members of the class, numbers 61264 and 61306, have been preserved.

No. 61016 *Inyala.* LNER 4-6-0 'Class B1' locomotive. Built at Darlington in January 1947 as LNER number 1016. She was given the name *Inyala* after a species of South African antelope (also spelt Nyala). Re-numbered 61016 by BR after nationalisation in 1948. Withdrawn from service in October 1965 whilst working out of York North Shed. Scrapped by Hughes Bolckows, North Blyth, in December 1965. (Photo P.68)

No. 61033 *Dibatag.* LNER 4-6-0 'Class B1' locomotive. Built at Darlington in August 1947 as LNER number 1033. She was given the name *Dibatag* after a species of South African antelope. Re-numbered 61033 by BR after nationalisation in 1948. Withdrawn from service in March 1963 whilst working out of Canklow Shed, Rotherham, South Yorkshire. Scrapped at Doncaster Works in March 1963. (Photo P.68)

No. 61038 *Blacktail.* LNER 4-6-0 'Class B1' locomotive. Built at Darlington for BR in November 1950 as BR number 61038. She was given the name *Blacktail* after a species of South African antelope. Withdrawn from service in May 1964 whilst working out of Gateshead Shed. Scrapped by Arnott Young of Dinsdale, Darlington in September 1964. (Photo P.18)

No. 61082. LNER 4-6-0 'Class B1' locomotive. Built in October 1946 by the North British Locomotive Works, Glasgow as LNER number 1082. She was un-named. Allocated BR number 61082 after nationalisation in 1948 whilst based at Immingham Shed. Withdrawn from service at Immingham Shed in December 1962 and scrapped at Cashmores, Great Bridge, Tipton, near Birmingham in February 1963. (Photo P.27)

No. 61145. LNER 4-6-0 'Class B1' locomotive. Built in June 1946 by the Vulcan Foundry as LNER number 1145. Allocated BR number 61145 after nationalisation in 1948 whilst based at Sheffield Darnall Shed. Withdrawn from service in January 1966 whilst working out of Colwick Shed, Nottingham. Scrapped at Cohens, Kettering in March 1966. (Photo P.110)

No. 61159. LNER 4-6-0 'Class B1' locomotive. Built in May 1947 by the Vulcan Foundry as LNER number 1159. Allocated BR number 61159 after nationalisation in 1948 whilst based at Gorton Shed in Manchester. Withdrawn from service in September 1963 whilst working out of Immingham Shed. Scrapped at Cashmores, Great Bridge, Tipton, West Midlands in November 1963. (Photo P.110)

No. 61179. LNER 4-6-0 'Class B1' locomotive. Built in June 1947 by the Vulcan Foundry as LNER number 1179. Allocated BR number 61179 after nationalisation in 1948 whilst based at Sheffield Darnall Shed. Withdrawn from service at Immingham Shed in January 1965 and scrapped at Drapers Yard, Hull in February 1965. (Photo P.20)

No. 61243 *Sir Harold Mitchell.* (Named after a former LNER Director.) LNER 4-6-0 'Class B1' locomotive. Built in October 1947 by the North British Locomotive Works, Glasgow as LNER number 1243. Allocated BR number 61243 in 1948. Withdrawn from service in May 1964 whilst based at Ayr Shed in Scotland. Scrapped by Arnott Young of Troon in July 1964. (Photo P.57)

No. 61247 *Lord Burghley.* LNER 'Class B1' locomotive. Built in October 1947 by the North British Locomotive Works, Glasgow as LNER number 1247. In December of that year, she was given the name *Lord Burghley* after a company director with the LNER. He was first appointed a director in 1934 and served until 1943 when he resigned. He was re-appointed in 1945 and remained in office until nationalisation of the railways in January 1948. Number 1247 was allocated BR number 61247 after nationalisation and she retained this until being withdrawn from service in May 1962, whilst working out of Colwick Shed, Nottingham. She was cut up for scrap at Doncaster Works in June 1962. (Photo P.20)

No. 61291. LNER 'Class B1' locomotive. Built for BR in February 1948 by the North British Locomotive Works, Glasgow as number 61291 and assigned to Darlington Shed. Withdrawn from service in May 1965 whilst working out of Ardsley Shed near Leeds. Scrapped in June 1965 by Ellis Metals, Swalwell, Gateshead. (Photo P.109)

No. 61306. LNER 'Class B1' locomotive. Built for BR in April 1948 by the North British Locomotive Works, Glasgow as number 61306 and assigned to Hull Dairycoates Shed. Withdrawn from service in September 1967 whilst working out of Low Moor Shed, Bradford. Number 61306 was subsequently preserved and given the name *Mayflower* whilst in preservation. She is now fully operational and can be seen on the North Norfolk Railway. Her owner is Mr David Buck. (Photo P.109)

No. 61319. LNER 'Class B1' locomotive. Built for BR in May 1948 by the North British Locomotive Works, Glasgow as number 61319 and assigned to Borough Gardens Shed in Gateshead. Withdrawn from service in December 1966 whilst working out of York North Shed. Scrapped in February 1967 by Drapers of Hull. (Photo P.67)

No. 61323. LNER 'Class B1' locomotive. Built for BR in May 1948 by the North British Locomotive Works, Glasgow as number 61323 and assigned to Kittybrewster Shed, Aberdeen. Withdrawn from service in November 1963 whilst working out of March Shed in Cambridgeshire. Scrapped in February 1964 at Doncaster Works. (Photo P.105)

LNER 'Class B16' (former NER 'Class S3') locomotives: general information

A total of seventy 4-6-0, NER 'Class S3', mixed traffic locomotives, designed by Vincent Raven, were built by the North Eastern Railway Company between 1919 and 1924. They were 3-cylinder mixed traffic locomotives, with a power classification of 6 MT – later re-classified 5MT by BR. The locomotives were absorbed into the stock of the LNER when the railway groupings took place in 1923 and they were re-classified as LNER 'Class B16' locomotives.

The LNER later rebuilt some of the 'B16' locomotives and decided to sub-classify them. The original Vincent Raven NER design engines which were not rebuilt were sub-classified as 'Class B16/1' locomotives.

Seven locomotives which were rebuilt by Nigel Gresley between 1937 and 1940 were sub-classified as 'Class B16/2' locomotives.

A further seventeen that were rebuilt by Edward Thompson between 1942 and 1949 were sub-classified as 'Class B16/3' locomotives.

No. 61421. LNER (NER) 4-6-0 'Class B16/2' locomotive. Built at Darlington in December 1920 as North Eastern Railway 'Class S3' locomotive number NER 936. Absorbed into the LNER in 1923 and re-classified as LNER 'Class B16' number 1421. This locomotive was rebuilt by Nigel Gresley in 1937 and re-classified as an LNER B16/2 locomotive. After nationalisation in 1948, received BR number 61421. Withdrawn from service whilst working out of York North Shed in June 1964, and scrapped by Hughes Bolckows, North Blyth, in December 1965. (Photo P.66)

No. 61441. LNER (NER) 4-6-0 'Class B16/1' locomotive. Built at Darlington in March 1923 as LNER number 1441. After nationalisation in 1948, received BR number 61441. Withdrawn from service in October 1959 whilst based at York North Shed. Scrapped at Darlington Works in November 1959. This locomotive retained her original Vincent Raven NER design throughout her service and was never rebuilt. (Photo P.66)

No. 61454. LNER (NER) 4-6-0 'Class B16/3' locomotive. Built at Darlington in October 1923 as LNER 'Class B16', number 1454. This locomotive was rebuilt by Edward Thompson in 1944 and re-classified as an LNER B16/3 locomotive. After nationalisation in 1948, received BR number 61454. Withdrawn from service whilst working out of York North Shed in June 1964, and scrapped by Hughes Bolckows, North Blyth, in September 1964. (Photo P.29)

No. 61455. LNER (NER) 4-6-0 'Class B16/2' locomotive. Built at Darlington in October 1923 as LNER 'Class B16', number 1455. This locomotive was rebuilt by Nigel Gresley in 1937 and re-classified as an LNER B16/2 locomotive. After nationalisation in 1948, received BR number 61455. Withdrawn from service whilst working out of York North Shed in September 1963 and scrapped at Darlington Works in October 1963.

LNER 4-6-0 'Class B17' locomotives: general information

A total of seventy-three 'Class B17' 3-cylinder locomotives, designed by Nigel Gresley, were built for the LNER between 1928 and 1937. These locomotives were designed as express passenger locomotives to replace the LNER (former GER) 'Class B12' locomotives which were no longer powerful enough to cope with heavier passenger trains being used on the former Great Eastern Railway main line from London to Cambridge, Ipswich and Norwich.

Gresley was unable to use his powerful passenger locomotives built for the East Coast Main Line due to weight restrictions imposed on the lines serving East Anglia. Consequently, he designed the 'Class B17', which was quite a powerful, yet lightweight 4-6-0 locomotive.

The first forty-eight class members were each named after country houses. The first locomotive, number 1600 (BR 61600), was called *Sandringham* (after the Queen's private residence, Sandringham House). This led to these engines being called 'Sandringham Class' locomotives. The next twenty-five locomotives were named after football clubs and were often referred to as footballers. This however was merely a nickname.

LNER 'Class B17' (Subdivisions) – 'Class B17' locomotives were sub-divided as follows:

- **'Class B17/1'.** The first forty-eight members of the class, named after country houses and fitted with GER design tenders, were classified as 'B17/1' locomotives.
- **'Class B17/2'.** These were 'B17/1' locomotives which received modifications to springs and axles resulting from cracks which appeared on some locomotive frames.
- **'Class B17/3'.** Locomotives with further modifications to springs, bogies and axle boxes.
- **'Class B17/4'.** The last twenty-five class members which were fitter with LNER tenders that were larger than the original GER design tenders fitted to the 'B17/1' engines.
- **'Class B17/5'.** This category was used for just two members of the class, numbers 61659 (pictured in this book) and 61670, which were rebuilt and fitted with streamlined casing similar to that used on LNER 'Class A4' locomotives. These locomotives then worked the prestigious '*The East Anglian*' express passenger train which operated from London to Norwich. The streamlined casing was removed from both engines in 1951.
- **'Class B17/6'.** After 1942 most 'Class B17' locomotives were rebuilt with new type 100A boilers which were used on 'Class B1' locomotives. These locomotives were then re-classified as 'Class B17/6' locomotives.
- **'Class B2'.** Between 1945 and 1949, Edward Thompson rebuilt ten 'Class B17' locomotives which he converted from 3-cylinder to 2-cylinder locomotives. These ten locomotives were then re-classified as LNER 'Class B2' locomotives.

No. 61630 *Tottenham Hotspur (Thoresby Park).* LNER 4-6-0 'Class B17/1' locomotive. Built at Darlington in April 1931 as LNER number 2830 and given the name *Thoresby Park,* after Thoresby Hall country estate at Newark in Nottinghamshire. Renamed *Tottenham Hotspur* in January 1938. Re-numbered as LNER 1630 in 1946. Rebuilt in December 1948 as a 'Class B17/6' locomotive, number BR 61630. Withdrawn from service in August 1958 whilst based at March Shed in Cambridgeshire and scrapped at Doncaster works later that month. (Photo P.67)

No. 61653 *Huddersfield Town.* LNER 4-6-0 'Class B17/4' locomotive. Built at Darlington in April 1936 as LNER number 2853 (LNER 1653 from 1946). After nationalisation in 1948, received BR number 61653. Rebuilt as a 'Class B17/6' in May 1954. Withdrawn from service in January 1960 whilst based at March Shed in Cambridgeshire. Scrapped in February 1960. (Photo P.30)

No. 61659 *East Anglian (Norwich City).* LNER 4-6-0 'Class B17/5' locomotive. Built at Darlington in June 1936 as an LNER 'Class B17' locomotive, number 2859 and given the name *Norwich City.* In September 1947, she was one of just two members of the class (the other being 2870), to be fitted with streamlined casing and side valances, similar to that used on LNER 'Class A4' locomotives, as a result of being selected to work a new express passenger train service '*The East Anglian*', which was being introduced on 27 September 1937 to work between London and Norwich. Upon completion of the modifications, number 2859 was re-numbered LNER 1659 and renamed *East Anglian,* before being re-classified as an LNER 'Class B17/5' locomotive.

In 1941, the side valances were removed. This was in accordance with instructions that side valances were to be removed from all LNER streamlined locomotives for ease of maintenance during the war years. Side valances were never re-fitted to LNER locomotives.

After nationalisation of the railways in 1948, number 1659 was allocated her BR number 61659, and in July 1949 she was re-fitted with a new type 100A boiler. After the refit, she was technically re-classified as a 'Class B17/6' locomotive, although in practice she was still referred to as a 'Class B17/5' locomotive, due to her unique streamlined appearance. In January 1951, the streamlined casing was removed entirely from number 61659 and her streamlined sister locomotive (BR number 61670) and both became 'Class B17/6' locomotives. Number 61659 was withdrawn from service in March 1960 whilst based at Lowestoft Shed. She was scrapped at Stratford Locomotive Works later that month. (Photo P.53)

No. 61666 *Nottingham Forest.* LNER 4-6-0 'Class B17/6' locomotive. Built by Robert Stephenson and Hawthorn Ltd, Darlington in February 1937 as LNER 'Class B17/4' locomotive, number 2866 (LNER 1666 from 1946). After nationalisation in 1948, she received BR number 61666. Rebuilt as a 'Class B17/6' in May 1954. Withdrawn from service in March 1960 whilst based at Stratford Shed in East London. Scrapped at Stratford Works in February 1960. (Photo P.30)

No. 61671 *Royal Sovereign (Manchester City).* LNER 4-6-0 'Class B2' locomotive. Built by Robert Stephenson and Hawthorn Ltd, Darlington in June 1937 as LNER 'Class B17' locomotive 2871. She was given the name *Manchester City,* a name that she carried until 1946.

In April 1946, she was renamed *Royal Sovereign* after she was selected to perform 'royal train duties' between London (Liverpool Street) and King's Lynn Station which served the Royal Residence at Sandringham. She was turned out in fully lined passenger livery and re-numbered LNER 1671.

In August 1948, number 1671 was rebuilt by Edward Thompson as a 'Class B2', 2-cylinder locomotive and given her BR number, 61671. She was withdrawn from service in September 1958 whilst based at

Cambridge Shed and scrapped at Stratford Works later that month. (Photo P.31)

LNER 2-6-2 'Class V4' locomotives: general information

Just two LNER 'Class V4' locomotives, designed by Nigel Gresley, were built for the LNER in 1941. These mixed traffic locomotives were the last locomotives to be designed by Gresley before he died in 1941. The first one built was LNER number 3401 which was named *Bantam Cock*. The second engine was number 3402 which was an un-named locomotive yet commonly referred to by the name *Bantam Hen*. Shortly after being built, both engines were sent to Scotland where they worked the West Highland Line from 1943 to 1949 before being replaced by 'Class B1' locomotives. The two engines remained in Scotland working mixed traffic train services, until they were withdrawn from service and scrapped in 1957, when their boilers were life expired.

No. 61700 *Bantam Cock*. LNER 2-6-2 'Class V4' locomotive. Built at Doncaster in February 1941 as LNER number 3401 *Bantam Cock*. Re-numbered LNER 1700 in 1946, followed by BR 61700 in 1948. Withdrawn from service in March 1957 whilst based at Aberdeen Ferryhill Shed. Scrapped at Kilmarnock Works later that month. (This locomotive appears alongside 'Class A4' locomotive number 60009 *Union of South Africa* in the same photograph.) (Photo P.15)

LNER 2-6-2. 'Class K1 & K2' – (GNR 'Class H2 & H3') locomotives: general information

A total of ten 2-6-2 mixed traffic (4MT) locomotives, designed by Nigel Gresley, were built for the GNR in 1912 as GNR 'Class H2' locomotives. It was soon discovered that the 4ft 8in (1.42m) diameter boilers were too small and the locomotives were not very successful. After they were absorbed into the LNER in 1923 they were re-classified as LNER 'Class K1' locomotives. All ten were rebuilt in the 1930s with larger, 5ft 6in (1.68m) diameter boilers, after which they were classified as LNER 'Class K2/1' engines. Their performance improved significantly.

Between 1914 and 1921, Gresley built a further sixty-five brand new locomotives with the larger boilers which he classified as GNR 'Class H3' (later LNER 'Class K2') locomotives.

A number of LNER 'Class K2' locomotives were assigned to work on the West Highland line in Scotland and many were named after Scottish lochs.

No. 61720. LNER 2-6-0 'Class K2/1' locomotive. She was the first member of the class when built at Doncaster in August 1912 as a GNR 'Class H1' locomotive, number GNR 1630. Absorbed into the LNER in 1923 as LNER 'Class K1' number 4630. After being rebuilt in the 1930s with a larger boiler, she was re-classified as an LNER 'Class K2/1'. She was re-numbered LNER 1720 in 1946, followed by BR 61720 in 1948. She was withdrawn from service in June 1956 and stored at Cowlairs, Glasgow until May 1957, when she was scrapped at Kilmarnock Works. (Photo P.22)

LNER (GNR) 2-6-0 'Class K3' locomotives: general information

A total of 193 'Class K3' locomotives were built between 1920 and 1937. They were 3-cylinder, mixed traffic locomotives designed for the Great Northern Railway by Nigel Gresley. They were a development of his earlier 'Class K2' locomotives and had a power classification of 6MT. Originally, ten of these 'Class K3' locomotives entered service on the GNR as GNR 'Class H4' locomotives and at the time had the largest boilers ever seen in Britain (6ft/1.83m) as well as being the first ever 3-cylinder 2-6-0 locomotives.

After the 1923 railway groupings took place, the ten GNR locomotives were absorbed into the LNER and re-classified as LNER 'Class K3' locomotives and a further 183 were built between 1923 and 1937. These locomotives were classified simply as 'Class K3' until 1947 when the class was divided as follows:

- **'Class K3/1'**. The original ten locomotives built for the GNR were re-classified as 'Class K3/1' locomotives.
- **'Class K3/2'**. The locomotives built by the LNER were re-classified as 'Class K3/2' engines.
- **'Class K3/3'**. A batch of locomotives fitted with the Westinghouse braking system were classified as 'Class K3/3' locomotives.
- **'Classes K4' 'K5', & 'K6'**. Referred to other 'K3' locomotives with various modifications.

A number of 'Class K3' locomotives were also fitted with former Great Northern design tenders.

No. 61863. LNER 2-6-0 'Class K5' locomotive. Built at Darlington Works in September 1925 as an LNER 'Class K3' locomotive, number 1863. She was rebuilt into a 'Class K5' locomotive by Edward Thompson in 1945. Following nationalisation in 1948, she was allocated BR number 61863 which she retained until being withdrawn from service in June 1960 whilst based at Stratford Shed in East London. She was scrapped at Doncaster Works at the end of June 1960.

Number 61863 was unique, due to her being the only LNER 'Class K5' locomotive ever built. She started life as an LNER 'Class K3', 3-cylinder locomotive designed by Nigel Gresley who was an advocate of 3-cylinder locomotives. His successor, Edward Thompson, however, was hostile towards 3-cylinder designs and supported the simpler 2-cylinder design. Out of a total of 193 'Class K3' locomotives produced, number 61893 was the only class member to be converted by Thompson into a 2-cylinder locomotive which he then re-classified as a 'Class K5' mixed traffic locomotive. (Photo P.22)

No. 61868. LNER 2-6-0 'Class K3' locomotive. Built at Darlington Works in October 1925 as LNER number 1868. Re-classified as an LNER 'Class K3/2' locomotive in 1947 and allocated BR number 61868 after 1948. Withdrawn from service in May 1962 whilst based at Doncaster Shed. Scrapped at Doncaster works later that month. (Photo P.11)

No. 61897. LNER 2-6-0 'Class K3' locomotive. Built at Darlington Works in August 1930 as LNER number 1897. Re-classified as an LNER 'Class K3/2' locomotive in 1947 and allocated BR number 61897 in 1948. Withdrawn from service in December 1962 whilst based at Hull Dairycoates Shed. Scrapped at Doncaster works in February 1963. (Photo P.104)

No. 61899. LNER 2-6-0 'Class K3' locomotive. Built by Armstrong Whitworth and Co., Newcastle in March 1931, as LNER number 1899. Re-classified as an LNER 'Class K3/2' locomotive in 1947 and allocated BR number 61899 in 1948. Withdrawn from service in December 1962 whilst based at Hull Dairycoates Shed. Scrapped at Doncaster works in January 1963. (Photo P.104)

No. 61935. LNER 2-6-0 'Class K3' locomotive. Built by Robert Stephenson & Hawthorn Ltd, Darlington, in December 1934 as LNER number 1935. Re-classified as an LNER 'Class K3/2' locomotive in 1947 and allocated BR number 61935 in 1948. Withdrawn from service in July 1962 whilst based at Hull Dairycoates Shed. Scrapped at Doncaster works in July 1962. (Photo P.13)

LNER 2-6-0 'Class K4' locomotives: general information

Just six LNER 'Class K4' locomotives, designed by Nigel Gresley, were built between 1937 and 1939, specifically to work services on the West Highland Line in Scotland. Due to the terrain of the line, Gresley had some basic problems to take into account when developing his new design. The West Highland Line had numerous sharp curves and steep gradients to navigate, as well as weight restrictions in place for locomotives using the line.

His solution was the 'Class K4' 3-cylinder locomotive, which was a powerful (6MT), yet a fairly light, 2-6-0 locomotive. To keep their weight to a minimum these engines were built with relatively small boilers and small driving wheels, measuring 5ft 2in (1.58m).

The 'Class K4' locomotives turned out to be a complete success and the uneconomical use of using small, double-headed locomotives on the line, which had previously been necessary due to weight restrictions prohibiting the use of large powerful locomotives, was removed. All six locomotives were given Scottish names.

No. 61995 *Cameron of Lochiel*. LNER 2-6-0 'Class K4' locomotive. Built at Darlington Works in December 1938 as LNER number 1995. Re-numbered BR 61995 in 1948. Withdrawn from service in October 1961 whilst based at Thornton Junction Shed, Fife in Scotland. She was scrapped at Halbeith Wagon Works in Dunfermline in January 1962. The locomotive was named after a Scottish West Highland clan chief. (Photo P.56)

LNER 2-6-0 'Class K1' locomotives: general information
A total of seventy 'Class K1', 2-6-0 'Mogul' locomotives designed by Edward Thompson and Arthur Peppercorn were built by the North British Locomotive Company in Glasgow between 1949 and 1950. These engines were simple 2-cylinder locomotives as opposed to the 3-cylinder designs preferred by Gresley. They were designed and built as mixed traffic locomotives (5P6F) and were the last steam locomotives of LNER design ever built for BR. One member of the class, number 62005, has been preserved.

No. 61997 *MacCailin Mor*. LNER 'Class K1/1' locomotive, built at Darlington in December 1938. Originally built as a 'Class K4' locomotive number LNER 3445 and given the name *MacCailin Mor*. In 1945, she was rebuilt, converting her from a 3-cylinder to a 2-cylinder locomotive and introducing various other modifications. This rebuilt locomotive was then re-classified as an LNER 'Class K1/1' and became the prototype for the new Thompson/Peppercorn 'Class K1' locomotives, of which seventy were built.

In 1948, LNER 3445 was allocated BR number 61997 and retained her name, *MacCailin Mor*. She was the only LNER 'Class K1/1' locomotive ever built. She was withdrawn from service in June 1961 whilst based at Stirling Shed, and scrapped at Doncaster Works later that month. (Photo P.51)

No. 62012. LNER 2-6-0 'Class K1' locomotive. Built in July 1949 by the North British Locomotive Works, Glasgow as BR number 62012. Withdrawn from service in May 1967 whilst based at Sunderland South Dock Shed. Scrapped by Drapers of Hull in August 1967. (Photo P.57)

No. 62052. LNER 2-6-0 'Class K1' locomotive. Built in November 1949 by the North British Locomotive Works, Glasgow as BR number 62052. Withdrawn from service in February 1963 whilst based at Fort William Shed in Scotland. Scrapped at Cowlairs Works, Glasgow in April 1964. She was the first member of the class to be withdrawn from service. (Photo P.57)

LNER (GNSR) 4-4-0 'Class D41' locomotives: general information
LNER 'Class D41' locomotives were a combination of two different classes of locomotive (GNSC 'Class S' and GNSC 'Class T'), which were originally designed and built as passenger locomotives (power classification 2P), for the Great North of Scotland Railway which became part of the LNER in 1923.

A total of six 'Class S' locomotives, designed by James Johnson, were introduced onto the GNSR in 1893, and a further twenty-six 'Class T' engines, designed by William Pickersgill, were built between 1895 and 1898. Both classes were manufactured by Neilson Reid and Company, Glasgow, and were almost identical in design apart from the fact that the Pickersgill design had a slightly larger boiler.

All thirty-two locomotives (the two classes combined) were absorbed into the LNER in 1923 and re-classified into a single class of engines, namely, LNER 'Class D41'. Withdrawals started in 1946 and although twenty-two survived into the BR era, their days were numbered and all were withdrawn from service and scrapped by 1953.

No. 62255. LNER (GNSC) 4-4-0 'Class D41' locomotive. Built in February 1898 by Neilson Reid and Co., Glasgow, as GNSR 'Class T' locomotive number GNSR 111 (later 111S). Absorbed into LNER in 1923 and re-classified as an LNER 'Class D41', number 2255. After nationalisation in 1948, became BR number 62255. Withdrawn from service in May 1952 whilst based at Keith in Moray, Scotland and later scrapped. (Photo P.92)

LNER (NER) 4-4-0 'Class D20' locomotives: general information

A total of sixty 'Class R' locomotives, designed by Wilson Worsdell, were built for the North Eastern Railway Company between 1899 and 1907. They were designed as express passenger locomotives, to be used for working main line services between York and Leeds and also from York to Newcastle and Edinburgh. With a power classification of 2P, these economical engines were very popular amongst the drivers and were the most successful express locomotives ever built for the NER.

All sixty 'Class R' locomotives were absorbed into the LNER in 1923 when they were re-classified as LNER 'Class D20' locomotives. In 1936, two members of the class were rebuilt by Nigel Gresley with long-travel piston valves and re-classified as 'Class D20/2' locomotives. The remaining fifty-eight were then re-classified as 'Class D20/1' locomotives.

Withdrawals started in 1943 but forty-six 'Class D20/1' locomotives and the two 'Class D20/2' engines survived to see BR service. Withdrawals continued through the 1950s and the whole class had been withdrawn from service by December 1957. All sixty were scrapped.

No. 62386. LNER (NER) 4-4-0 'Class D20/1' locomotive. Built in August 1907 at Gateshead North Eastern Railway Locomotive Works as NER 'Class R' locomotive number 1207. Absorbed into stock of the LNER after 1923 and re-classified as LNER 'Class D20', number 2386. Allocated BR number 62386 in 1948 (displayed from June 1950). Withdrawn from service in October 1956 whilst based at Selby, where she had spent her entire BR service, before being transported to Darlington Works and scrapped. (Photo P.92)

LNER (GER) 4-4-0. 'Class D15' (Also 'Classes D14' & 'D16') locos: general information

Three very similar classes of express passenger locomotives were built in sets of batches for the Great Eastern Railway between 1900 and 1923 (different batches had various modifications). They were GER 'Class S46', 'Class D56' and 'Class H88'. The first batch of these engines was the 'Class S46'. A total of forty-one of these locomotives (introduced in five sets), designed by James Holden, were built between 1900 and 1903. The second batch (seven sets), consisting of seventy 'Class D56' locomotives, also designed by James Holden, was built between 1903 and 1911. The final batch consisted of just one set of ten 'Class H88' locomotives, designed by Alfred John Hill, and was built by the LNER in 1923, after the railway groupings had taken place.

The first of these locomotives to be built, and the only one to be named, was number GER 1900 ('Class S46'). She was given the name *Claud Hamilton*, after Lord Claud John Hamilton who was a prominent MP and Chairman of the Great Eastern Railway Company at the time. Locomotives from each of the three classes were subsequently referred to as Claud Hamilton Class locomotives. The 'Class H88' locomotives, which were fitted with larger superheated Belpaire boilers, were also referred to as 'Super Clauds'.

Many of the above-mentioned locomotives were rebuilt and were later re-classified by the LNER as follows:

GER 'Class S46' locomotives were classified as LNER 'Class D14' locomotives.

GER 'Class D56' locomotives were classified as LNER 'Class D15' locomotives. The 'Class D15' engines were further sub-divided as follows:

- 'Class D15' locomotives with short smokeboxes were classified as 'Class D15/1' engines.
- 'Class D15' locomotives with long smokeboxes were classified as 'Class D15/2' engines.

GER 'Class H88' locomotives (Super Clauds) were classified as 'Class D16' locomotives.

- 'Class D16' locomotives with short smokeboxes were classified as 'Class D16/1' engines.
- 'Class D16' locomotives with long smokeboxes were classified as 'Class D16/2' engines.

Some other 'Class D15' and 'Class D16' engines which were later rebuilt by Nigel Gresley were re-classified as LNER **'Class D16/3'** locomotives.

Although all the above-mentioned locomotives were built with a power classification of 2P, the D16/3 locomotives were re-classified 3P by BR in 1953. During the BR era, these express passenger locomotives were frequently used for working goods and freight services.

No. 62503. LNER (GER) 4-4-0 'Class D15/2' locomotive. Built in April 1900 at Stratford Works in East London as Great Eastern Railway 'Class S46' number 1892. Absorbed into the LNER in 1923 and re-classified as LNER 'Class D15' locomotive number 8892 (2503 in 1946). Allocated BR number 62503 in 1948. Withdrawn from service from Bury St. Edmunds Shed in February 1951 and scrapped at Stratford Works later that month. (Photo P.91)

No. 62543. LNER (GER) 4-4-0 'Class D16/3' locomotive. Built in December 1903 at Stratford Works in East London as Great Eastern Railway 'Class D56' number 1852. Absorbed into the LNER in 1923 and re-classified as LNER 'Class D15' locomotive number 8852 (2543 in 1946). Rebuilt by Nigel Gresley as an LNER 'Class D16/3' engine, before being allocated BR number 62543 in 1948. Withdrawn from service from March Shed, Cambridgeshire in October 1958 and scrapped at Stratford Works in December 1958. (Photo P.81)

No. 62562. LNER (GER) 4-4-0 'Class D16/3' locomotive. Built in March 1908 at Stratford Works in East London as Great Eastern Railway 'Class D56' number 1831. Absorbed into the LNER in 1923 and re-classified as LNER 'Class D15' locomotive number 8831 (2562 in 1946). Rebuilt by Nigel Gresley as an LNER 'Class D16/3' engine, before being allocated BR number 62562 in 1948. Withdrawn from service from March Shed, Cambridgeshire in October 1957 and scrapped at Stratford Works in November 1957. (Photo P.28)

LNER (GCR) 4-4-0 'Class D10' 'Director Class' locomotives: general information

LNER 'Class D10'. A total of ten GCR 'Class 11E', express passenger locomotives (power classification 3P), designed by John Robinson, were built in 1913 to replace the earlier 'Class 11B' locomotives which were too small to cope with growing demands of the day. The ten locomotives were originally named after ten of the twelve directors of the Great Central Railway Company and were referred to as 'Director Class' locomotives. They were numbered GCR 429 to 438 and soon proved themselves to be very successful. A plan to build additional 'Class 11E' engines was suspended after the outbreak of war in 1914 and the order was later cancelled. The ten 'Class 11E' locomotives were absorbed into the LNER in 1923 when they were re-classified as LNER 'Class D10' locomotives.

All ten 'Class 11E' (LNER 'Class D10') 'Director Class' locomotives survived into the BR era and were withdrawn from service between 1953 and 1955. They were all scrapped.

No. 62650 *Prince Henry.* LNER (GCR) 4-4-0 'Class D10' locomotive. Built in August 1913 at Gorton Locomotive Works, Manchester, as Great Central Railway 'Class 11E' locomotive number 429. Absorbed into the LNER in 1923 and re-classified as LNER 'Class D10' locomotive number 2650 (later BR 62650). Withdrawn from service in February 1954 whilst based at Northwich Shed, Cheshire, and scrapped at Gorton Works later that month.

Number 62650 was originally named *Sir Alexander Henderson* (Director of the GCR) when built. The name was changed in 1917 to *Sir Douglas Haig* (British Army field marshal during the First World War). The name was changed a second time in 1920 to *Prince Henry* and remained on the locomotive until she was withdrawn from service. Prince Henry (Duke of Gloucester) was the third son (fourth child) of King George V and Queen Mary. (Photo P.80)

No. 62653 *Sir Edward Fraser.* LNER (GCR) 4-4-0 'Class D10' locomotive. Built in October 1913 at Gorton Locomotive Works, Manchester, as Great Central Railway 'Class 11E' locomotive number 432. Absorbed into the LNER in 1923 and re-classified as LNER 'Class D10' locomotive number 2653 (BR 62653 after 1948). Withdrawn from service in October 1955 whilst based at Northwich Shed, Cheshire, and scrapped at Gorton Works later that month. She was the last member of the class to be withdrawn from service. (Photo P.42)

No. 62654 *Walter Burgh Gair.* LNER (GCR) 4-4-0 'Class D10' locomotive. Built in October 1913 at Gorton Locomotive Works, Manchester, as Great Central Railway 'Class 11E' locomotive number 433. Absorbed into the LNER in 1923 and re-classified as LNER 'Class D10' locomotive number 2654 (BR 62654 after 1948). Withdrawn from service in September 1953 whilst based at Trafford Park Shed, Manchester and scrapped at Gorton Works later that month. (Photo P.42)

LNER (GCR) 4-4-0 'Class D11' 'Improved Directors Class': general information

Initially, eleven GCR 'Class 11F' locomotives were built for the GCR between 1919 and 1920. These engines were an improved version of the GCR 'Class 11E' (LNER 'Class D10') 'Director Class' locomotives and were referred to as 'Improved Directors'. They were absorbed into the LNER in 1923 and re-classified as LNER 'Class D11' locomotives.

A further twenty-four 'Class 11F' (LNER 'Class D11') locomotives were built for the LNER in 1924 for use in Scotland and they were referred to as 'Scottish Directors'. Each were named after Sir Walter Scott novels.

The LNER 'Class D11' locomotives were sub-divided as follows:

- **'D11/1'.** The eleven locomotives built between 1919 and 1920 became LNER 'Class D11/1'.
- **'D11/2'.** The twenty-four 'Scottish Directors' built in 1924 became LNER 'Class D11/2'.

All thirty-five 'Class D11' locomotives were withdrawn from service between 1958 and 1962.

No. 62666 *Zeebrugge.* LNER (GCR) 4-4-0 'Class D11/1' locomotive. Built in October 1922 as Great Central Railway 'Class 11F' locomotive number 502. Absorbed into the LNER in 1923 and re-classified as LNER

'Class D11' number 5502 (LNER 2666 in 1946). Entered BR service in 1948 as BR number 62666. Withdrawn from service in December 1960 whilst based at Sheffield Darnall Shed and scrapped at Doncaster Works later that month. (Photo P.81)

LNER 4-4-0 'Class D49' 'Shire' & 'Hunt Class' locomotives: general information

A total of seventy-six LNER 'Class D49' locomotives, designed by Nigel Gresley, were built between 1927 and 1935. These 3-cylinder locomotives were designed to work ordinary passenger train services (power classification 4P). They were all built at Darlington Works.

The first thirty-six members of the class, built between 1927 and 1929, were named after counties and became known as 'Shire Class' locomotives. The remaining members of the class were named after fox hunts and were referred to as 'Hunt Class' locomotives. 'Class D49' locomotives were later sub-divided as follows:

The thirty-six 'Shire Class' engines (with the exception of two), became 'Class D49/1' engines. Some locomotives fitted with Lentz rotary-cam poppet valves became 'Class D49/2' engines. Other locomotives fitted with Lentz oscillating-cam valves became 'Class D49/3' engines.

The whole class were withdrawn from service between 1957 and 1961. Just one member of the class, number 62712 *Morayshire,* has been preserved and can be seen on the Bo'ness and Kinneil Heritage Railway in Scotland.

No. 62706. *Forfarshire.* LNER 4-4-0 'Class D49/1' locomotive. Built at Darlington Works in December 1927 as LNER number 2706. Re-numbered 62706 by BR in 1948. Withdrawn from service in February 1958 whilst based at Thornton Junction Shed in Fife, Scotland and scrapped at Darlington Works later that month. (Photo P.93)

No. 62726 *The Meynell.* LNER 4-4-0 'Class D49/2' locomotive. Built at Darlington Works in March 1929 as LNER number 2726. Re-numbered 62726 by BR in 1948. Withdrawn from service in December 1957 whilst based at Scarborough Shed and scrapped at Darlington Works in March 1958.

Number 62726 was given the name *Leicestershire* when she was built. The name was changed to *The Meynell* (named after a Staffordshire hunt) in 1932. In addition, she started life as a 'Class D49/1' locomotive but was later fitted with Lenz rotary cam poppet valves and re-classified as a 'Class D49/2' locomotive. (Photo P.100)

No. 62737 *The York and Ainsty.* LNER 4-4-0 'Class D49/2' locomotive. Built at Darlington Works in May 1932 as LNER number 2737. Re-numbered 62737 by BR in 1948. Withdrawn from service in January 1958 whilst based at Hull Botanic Gardens Shed, and scrapped at Darlington Works in February 1958. Number 62737 started life as a 'Class D49/1' locomotive but was later fitted with Lenz rotary cam poppet valves and re-classified as a 'Class D49/2' locomotive. (Photo P.82)

No. 62740 *The Bedale.* LNER 4-4-0 'Class D49/2' locomotive. Built at Darlington Works in June 1932 as LNER number 2740. Re-numbered 62740 by BR in 1948. Withdrawn from service in August 1960 whilst based at Hull Dairycoates Shed, and scrapped at Darlington Works in September 1960. Number 62740 started life as a 'Class D49/1' locomotive but was later fitted with Lenz rotary cam poppet valves and re-classified as a 'Class D49/2' locomotive. (Photo P.82)

No. 62756 The Brocklesby. LNER 4-4-0 'Class D49/2' locomotive. Built at Darlington Works in August 1934 as a Class D49/1 and later rebuilt as a 'Class D49/2'. Given the name 'The Brocklesby' after a Hunt at Beelsby near Grimsby in North Lincolnshire. Withdrawn from service whilst based at Scarborough in April 1958 and scrapped at Darlington later that month. Her boiler, still in good condition was sent to Cowlais (Glasgow) as a spare. (Photo P.82)

No. 62762 *The Fernie.* LNER 4-4-0 'Class D49/2' locomotive. Built at Darlington Works in September 1934 as LNER number 2762. Re-numbered 62762 by BR in 1948. Withdrawn from service in October 1960 whilst based at Scarborough Shed, and scrapped at Darlington Works in November 1960. Number 62762 started life as a 'Class D49/1' locomotive but was later fitted with Lenz rotary cam poppet valves and re-classified as 'Class D49/2'. (Photo P.100)

LNER (GER) 2-4-0 'Class E4' locomotives: general information

A total of 100 'Class T26' locomotives, designed by James Holden as mixed traffic locomotives (power classification 1MT), were built at Stratford Locomotive Works for the Great Eastern Railway between 1891 and 1902. They were absorbed into the LNER in 1923 and re-classified as LNER 'Class E4' locomotives. Eighty-two were withdrawn from service and scrapped between 1926 and 1940. The remaining eighteen went on to serve with BR but were all withdrawn in the 1950s. Number 62785 was preserved as a part of the national collection and the remainder were scrapped.

No. 62785. LNER (GER) 2-4-0 'Class E4' locomotive. Built at Stratford Works in January 1895 as a Great Eastern Railway 'Class T26' locomotive, number 490. Absorbed into the LNER in 1923 and re-classified as LNER 'Class E4' locomotive number 7490, later 7802 and finally 2785. Allocated BR number 62785 in 1948. Withdrawn from service in December 1959 whilst based at Cambridge Shed, when she was the last 2-4-0 tender locomotive in service on British Railways.

After being withdrawn from service, number 62785 was preserved and restored in her original GER livery and displaying her GER number, 490. She was displayed in the British Transport Museum at Clapham in London before moving to the National Railway Museum in York. She is now a part of the national collection of steam locomotives but is currently on loan from York Museum and is on public display at Bressingham Steam Museum in Norfolk. (Photo P.96)

No. 62793. LNER (GER) 2-4-0 'Class E4' locomotive. Built at Stratford Works in June 1902 as a Great Eastern Railway 'Class T26' locomotive number 408. Absorbed into the LNER in 1923 and re-classified as LNER 'Class E4' locomotive number 7408, later 2793. Allocated BR number 62793 in 1948. Withdrawn from service in February 1955 whilst based at Norwich Thorpe Shed. Scrapped at Stratford Works in August 1955. (Photo P.96)

LNER (NER) 0-8-0 'Class Q5' locomotives: general information

A total of forty NER 'Class T' locomotives and fifty 'Class T1' locomotives, designed by Wilson Worsdell, were built for the North Eastern Railway Company between 1901 and 1911. The difference between the two classes was that the 'Class T' locomotives had piston valves, whilst the 'Class T1' locomotives had slide valves. These freight locomotives (power classification 6F) were designed exclusively for working heavy mineral traffic.

When the two classes of locomotive were absorbed into the LNER in 1923, they were both re-classified in the same class, and all ninety became LNER 'Class Q5' locomotives. Fourteen of these locomotives were rebuilt with larger boilers between 1932 and 1934. These rebuilds were then re-classified as LNER 'Class Q5/2' locomotives and the remaining un-rebuilt locomotives were re-classified as 'Class Q5/1'.

Withdrawals started in 1946 and although seventy-seven entered BR service in 1948, their days were numbered and they were all withdrawn for scrap before the end of 1951.

No. 63326. LNER (NER) 0-8-0 'Class Q5/1' locomotive. Built at Darlington Works in July 1911 as North Eastern Railway

'Class T' locomotive number 669. Absorbed into LNER in 1923 as LNER 'Class Q5' number 3326. Allocated BR number 63326 in 1948. Withdrawn from service in October 1951 whilst based at Borough Gardens Shed, Gateshead, and scrapped at Darlington works shortly afterwards. Although number 63326 was allocated her BR number in 1948 she never wore it. She still bore her LNER number 3326 when she was scrapped. (Photo P.56)

LNER (NER) 0-8-0 'Class Q6' locomotives: general information

A total of 120 North Eastern Railway 'Class T2' locomotives, designed by Vincent Raven, were built between 1913 and 1921. These heavy freight locomotives were based on the designs of the NER 'Class T' and 'Class T1' (LNER Class Q5), mentioned earlier. After being absorbed into the LNER in 1923, they were re-classified as LNER 'Class Q6'.

These engines were very successful freight locomotives and could be seen operating in the north east of England until the very last days of steam. The last member of the class was withdrawn from service in 1967.

No. 63395. LNER (NER) 0-8-0 'Class Q6' locomotive. Built at Darlington for the North Eastern Railway Company in December 1918 as NER 'Class T2' locomotive number 2238. Absorbed into the LNER in 1923 and re-classified as an LNER 'Class Q6', number 3395. Allocated BR number 63395 in 1948. Withdrawn from service in September 1967 whilst based at Sunderland South Dock Shed.

After being withdrawn from service, number 63395 was bought by the North Eastern Locomotive Preservation Group for restoration and preservation. They still own her, and she can be seen restored to her former glory and working on the North Yorkshire Moors Heritage Railway where she is based. (Photo P.47)

No. 63406. LNER (NER) 0-8-0 'Class Q6' locomotive. Built at Darlington Works in June 1919 as North Eastern Railway 'Class T2' locomotive number 2249. Absorbed into LNER in 1923 as LNER 'Class Q5' number 3406. Allocated BR number 63406 in 1948. Withdrawn from service in July 1966 whilst based at Sunderland South Dock Shed, and scrapped at Hughes Bolckows, Blyth in December 1966. (Photo P.70)

LNER 'Class O4' (Former GCR 'Class 8K'): general information

A total of 126 GCR 'Class 8K' 2-8-0 heavy freight locomotives, designed by John Robinson, were built between 1911 and the outbreak of the First World War in 1914. The locomotive building programme for the GCR 'Class 8K' was suspended during the war years but continued after the war in 1918.

During the actual war years, a further 500 of these 2-8-0 locomotives were built for the British Government to be used by the Railway Operating Division (ROD) of the British Army (Royal Engineers), for wartime requirements home and abroad.

In 1923, the LNER inherited 131 'Class 8K' locomotives from the GCR and re-classified them as LNER 'Class O4'. The LNER later purchased 273 similar locomotives from the ROD which had been built for wartime use. These were also classified as LNER 'Class O4' locomotives.

The locomotives were later sub-divided as follows:

'Class O4/1'. Original GCR 'Class 8K' engines built between 1911 and 1914.
'Class O4/2'. ROD engines with reduced cab and boiler height for loading gauge.
'Class O4/3'. ROD engines with steam brake only (no vacuum brake) and no water scoop. All the ROD 'Class O4/3' locomotives were all re-classified as 'Class O4/1' in 1947.
'Class O4/4'. Used for 'Class 8K' engines which were rebuilt with 'O2 boilers'.
'Class O4/5'. Engines rebuilt after 1932 with boiler and firebox modifications.
'Class O4/6'. Originally LNER 'Class O5' locomotives rebuilt with smaller boilers.

'Class O4/7'. Engines rebuilt after 1939 with similar modification to the 'Class O4/5'.

'Class O4/8'. Engines rebuilt after 1944 with 'Class B1' type boilers and side-window cab.

No. 63570. LNER (GCR) 2-8-0 'Class O4/7' locomotive. Built in October 1912 by the North British Locomotive Works, Glasgow as a Great Central Railway 'Class 8K' locomotive number 1223. Absorbed into the LNER in 1923 as LNER number 6223 (3570 in 1946) and BR 63570 in 1948. Number 63570 was rebuilt by Nigel Gresley after 1939. She was withdrawn from service in December 1961 whilst based at Ardsley Shed near Leeds and scrapped at Gorton Works, Manchester in February 1962. (Photo P.55)

No. 63857. LNER (GCR) 2-8-0 'Class O4/7' locomotive. Built in July 1919 by the North British Locomotive Works, Glasgow as a Great Central Railway 'Class 8K' locomotive number 1618. Absorbed into the LNER in 1923 as LNER number 6618 (3857 in 1946) and BR 63857 from 1948. The locomotive was rebuilt by Nigel Gresley after 1939. Withdrawn from service in August 1962 and scrapped in November 1962. (Photo P.71)

No. 63893. LNER (GCR) 2-8-0 'Class O4/8' locomotive. Built in June 1919 by the North British Locomotive Works, Glasgow as a Great Central Railway 'Class 8K' locomotive number 1293. Absorbed into the LNER in 1923 as LNER number 6293 (3893 in 1946) and BR 63893 from 1948. The locomotive was rebuilt by Edward Thompson after 1944. Withdrawn from service in June 1965 and scrapped in August 1965. (Photo P.71)

No. 63906. LNER (GCR) 2-8-0 'Class O4/6' locomotive. Built in June 1918 at the Gorton Locomotive Works, Manchester as a Great Central Railway 'Class 8K' locomotive number 1417. Absorbed into the LNER in 1923 as LNER number 6417 (3906 in 1946) and BR 63906 from 1948. Withdrawn from service in January 1965 and scrapped in April 1965. (Photo P.72)

LNER 'Class J6', GNR 'Classes 521 & 536' (GNR J22): general information

A total of 110 Great Northern Railway 0-6-0 'Class 536' locomotives, designed by Henry Ivatt and Nigel Gresley, were built at Doncaster Works between 1911 and 1922. The first fifteen were built in 1911, to the design of Ivatt ('Class 521'). The remainder were slightly modified by Gresley ('Class 536') and built between 1912 and 1922.

They were designed primarily as light goods locomotives (power classification 3F) but were soon found to be very versatile engines and were frequently used to work passenger train services. As such, they became used for general mixed traffic duties (2P3F).

The GNR classes 521 and 536 were absorbed into the LNER in 1923 as one single class, GNR 'Class J22', before being re-classified as LNER 'Class J6' locomotives.

All 110 locomotives passed to British Railways in 1948 and were withdrawn from service between 1955 and 1962. None was preserved.

No. 64170. LNER (GNR) 0-6-0 'Class J6' locomotive. Built at Doncaster in August 1911 as Great Northern Railway '521 Class' (J22) locomotive number 521. Absorbed into the LNER in 1923 and re-classified as LNER 'Class J6' locomotive number 3521 (4170 in 1946). BR number 64170 from 1948. Withdrawn from service in July 1961 whilst based at Ardsley Shed in West Yorkshire, and scrapped at Doncaster Works later that month. Number 64170 was the first member of the class to be built. (Photo P.95)

No. 64203. LNER (GNR) 0-6-0 'Class J6' locomotive. Built at Doncaster in June 1913 as Great Northern Railway '536 Class' (J22) locomotive number 554. Absorbed into the LNER in 1923 and re-classified as LNER 'Class J6' locomotive number 3554 (later LNER

4203). Allocated BR number 64203 in 1948. Withdrawn from service in June 1962 whilst based at Ardsley Shed in West Yorkshire, and scrapped at Doncaster Works later that month. (Photo P.95)

No. 64205. LNER (GNR) 0-6-0 'Class J6' locomotive. Built at Doncaster in June 1913 as Great Northern Railway '536 Class' (J22) locomotive number 556. Absorbed into the LNER in 1923 and re-classified as LNER 'Class J6' locomotive number 3556 (later LNER 4205). Allocated BR number 64205 in 1948. Withdrawn from service in October 1958 whilst based at Ardsley Shed in West Yorkshire, and scrapped at Doncaster Works in January 1959. (Photo P.54)

LNER 'Class J11', GCR 'Class 9J' 0-6-0 locomotive: general information

A total of 174 Great Central Railway 0-6-0 'Class 9J' locomotives, designed by John Robinson, were built for the GCR between 1901 and 1910. They were designed to work goods trains on the GCR and had a power classification of 3F. These locomotives were given the nickname 'Pom-Poms' due to the distinct noise made by their exhaust beats. During the First World War, eighteen 'Class 9J' locomotives were loaned to the Railway Operating Division and shipped to France for use in the conflict. They were all safely returned to Britain in 1919 after the war had ended.

The 'Class 9J' locomotives were absorbed into the LNER in 1923 and re-classified as LNER 'Class J11' locomotives. Whilst under LNER ownership, a variety of modifications were carried out, after which the class was divided into five different sub-classes as follows:

- **'Class J11/1'.** Locomotives fitted with 3,250 gallon tenders. High boiler mountings. Some superheated.
- **'Class J11/2'.** Locomotives fitted with 4,000 gallon tenders. High boiler mountings. Some superheated.
- **'Class J11/3'.** Thirty-one locomotives rebuilt with long travel piston valves.
- **'Class J11/4'.** Locomotives fitted with 3,250 gallon tenders. Low boiler mountings. All superheated.
- **'Class J11/5'.** Locomotives fitted with 4,000 gallon tenders. Low boiler mountings. All superheated.

Alterations continued to be made to 'Class J11' locomotives until in 1952 they had all superheated boilers and were fitted with low boiler mountings. This led to the various sub-classes being abolished and the whole class once again classified as 'Class J11'.

Although all the 'Class J11' engines entered BR service in 1948, withdrawals started in 1954 and continued until 1962. The whole class was scrapped.

No. 64417. LNER (GCR) 0-6-0 'Class J11/3' locomotive. Built at Gorton Locomotive Works, Manchester in August 1907 as a Great Central Railway 'Class 9J' locomotive. Absorbed into the LNER in 1923 as a 'Class J11' (later re-classified as an LNER 'Class J11/3'). Allocated LNER number 4417 (later BR 64417). Withdrawn from service in August 1961 whilst based at Staveley Shed (later called Barrow Hill) in Derbyshire. Scrapped at Gorton Works in September 1961. (Photo P.85)

LNER 'Class J37' NBR 'Class S' 0-6-0 locomotives: general information

A total of 104 'Class S' locomotives, designed by William Paton Reid, were built for the North British Railway between 1914 and 1921. They were a superheated development of his 'Class J35' locomotive, designed and intended for use as express freight locomotives (power classification NBR 4F, LNER 4F, BR 5F). They were, however, later used as both passenger and goods locomotives as well as coal trains. They were very successful locomotives.

After being absorbed into the LNER, the NBR 'Class S' engines were re-classified as LNER 'Class J37' locomotives. The whole class survived the LNER era and saw service with BR from 1948 onwards. Most worked through the 1950s and although withdrawals started in 1959, many continued to work until the last one was finally withdrawn from service in 1966. Unfortunately, they were all scrapped and none was preserved.

No. 64639. LNER (NBR) 0-6-0 'Class J37' locomotive. Built at Cowlairs Works, Glasgow in October 1921 as a North British Railway 'S Class' locomotive, number 287. Absorbed into the LNER in 1923 and re-classified as an LNER 'Class J37', number 518 (later LNER 4639 and BR 64639). Withdrawn from service in December 1961 whilst based at Eastfield Shed, Glasgow, and scrapped at Barnes and Bell Scrapyard, Coatbridge, Glasgow in February 1962. Number 64639 was the last member of the class to be built. (Photo P.85)

LNER 'Class J39' 0-6-0 locomotives: general information

A total of 289 of these medium powered (4P5F), 0-6-0 locomotives, designed by Nigel Gresley, were introduced between 1926 and 1941. They could be seen operating all over the LNER network working as mixed traffic locomotives. Gresley based his 'Class J39' design on his 'Class J38' (6F), freight locomotives (also introduced in 1926) but with larger wheels and a shorter boiler. 'Class J39' locomotives were fitted with three different types of tender and as such, the class was sub-divided as follows:

Class J39/1. Locomotives fitted with standard LNER 3,500 gallon tenders.
Class J39/2. Locomotives fitted with standard LNER 4,200 gallon tenders.
Class J39/3. Locomotives fitted with former NER tenders from withdrawn locomotives.

All 'Class J39' locomotives were withdrawn from service and scrapped between 1959 and 1962, and none was preserved.

No. 64721. LNER 0-6-0 'Class J39/1' locomotive. Built at Darlington in May 1927 as LNER number 4721 (later BR 64721). Withdrawn from service in February 1960 whilst based at Doncaster Shed. Scrapped at Stratford Works the following month. (Photo P.48)

No. 64760. LNER 0-6-0 'Class J39/1' locomotive. Built at Darlington in September 1928 as LNER number 4760 (later BR 64760). Withdrawn from service in November 1962 whilst based at Ardsley Shed in West Yorkshire. Scrapped at Darlington Works in September 1963. (Photo P.47)

LNER 'Class J16' & 'Class J17', GER 'Class F48' & 'Class G58': general information

A total of sixty Great Eastern Railway 'Class F48', 0-6-0 tender locomotives, designed by James Holden, were built at Stratford Works between 1900 and 1903. They were built exclusively as freight locomotives (power classification 4F).

Holden improved the design of his 'Class F48' engines and between 1905 and 1911 he built a further thirty brand new locomotives which he classified as GER 'Class G58'. He then set about rebuilding all the 'Class F48' locomotives into 'Class G58' locomotives.

In 1923, the GER was absorbed into the LNER who re-classified the GER 'Class F48' engines as LNER 'Class J16', and the 'Class G58' engines as LNER 'Class J17'. The programme of rebuilding the 'F48' (LNER 'Class J16') locomotives into 'G58' (LNER Class J17) locomotives continued until all the 'Class J16s' had been rebuilt, thus increasing the total of 'Class J17' locomotives to ninety. LNER 'Class J16' engines then became extinct.

All the 'Class J17' locomotives were used successfully by the LNER and with the exception of one, they all entered BR service in 1948. They were withdrawn from service between 1953 and 1962. Just one member of the class, number 65567, was preserved by the National Railway Museum as a part of the national collection and is currently on public display at Barrow Hill Roundhouse.

The locomotive which did not see BR service was LNER number 8200, as it was destroyed in 1944 by a V2 rocket at Stratford, East London.

No. 65557. (GER) LNER 0-6-0 'Class J17' locomotive. Built at Stratford Works in February 1903 as Great Eastern Railway 'Class F48' locomotive number GER 1207. Rebuilt as a GER 'Class G58' in December 1921. Absorbed into the LNER in 1923 and re-classified as an LNER 'Class J17' number 8207. Re-numbered LNER 5557 in 1946, followed by BR number 65557 in 1948. Withdrawn from service in April 1959 whilst based at Melton Constable Shed, Norfolk, and scrapped at Stratford Works the following month. (Photo P.48)

LNER 'Class F6', GER 'Class G69' 2-4-2 tank locomotive: general information

A total of twenty, brand new GER 'Class G69' tank locomotives, designed by James Holden, were produced between 1911 and 1912. A further two members of the class were rebuilt from the earlier GER 'Class M15' engines which had been designed by his father, James Holden, in 1904. These 2-4-2 tank engines were built for light passenger work and they were particularly useful for working on branch lines. Some of the engines were fitted with push and pull apparatus for this purpose.

In January 1923, the 'Class G69' locomotives were absorbed into the LNER and re-classified as LNER 'Class F6' locomotives. The whole class was then assigned to Stratford Shed in East London from where they worked London commuter trains from Liverpool Street and Fenchurch Street to Southend and Tilbury. All twenty-two 'Class F6' locomotives entered BR service in 1948 although their working life was coming to an end. Withdrawals started the following year and the last five members of the class were withdrawn from service in 1958. They were all scrapped.

No. 67238. LNER (GER) 2-4-2T 'Class F6' tank locomotive. Built at Stratford Works in January 1912 as a Great Eastern Railway, 'Class G69' tank locomotive number GER number 9. Absorbed into the LNER in 1923 and re-classified as an LNER 'Class F6' locomotive number 7009. Re-numbered LNER 7238 in 1946, followed by BR number 67238 in 1948. Withdrawn from service in November 1955 whilst based at Cambridge Shed, and scrapped at Stratford Works in January 1956. (Photo P.64)

LNER 'Class C12', GNR 'Class C2' 4-4-2 tank locomotive: general information

A total of sixty 'Class C2' tank locomotives, designed by Henry Ivatt, were built for the Great Northern Railway between 1898 and 1907. After being absorbed into the LNER in 1923, they were re-classified as LNER 'Class C12'. They were designed and built to work local passenger train services in north east England and commuter services in north London. Withdrawals started in 1937 but forty-nine entered BR service in 1948. All were withdrawn from service in the 1950s and 1958 saw the last few being withdrawn for scrap. None was preserved.

No. 67356. LNER (GNR) 4-4-2 'Class C12' tank locomotive. Built at Doncaster Works in December 1900 as GNR 'Class C2' locomotive number GNR 1018. Absorbed into the LNER in 1923 and re-classified as LNER 'Class C12' locomotive, number 4018. Re-numbered LNER 7356 in 1946, followed by BR number 67356 in 1948. Withdrawn from service in October 1951, whilst based at King's Lynn Town Shed. Scrapped at Doncaster Works later that month. (Photo P.64)

LNER 'Class C15', NBR 'Class M' 4-4-2 tank locomotives: general information

A total of thirty 4-4-2 NBR 'Class M' tank locomotives, designed by William Paton Reid, were built by the Yorkshire Engine Company, Sheffield, for the North British Railway Company between 1911 and

1913. They were originally designed and built to operate suburban passenger trains around Edinburgh and Glasgow. They also worked passenger stopping trains between the two cities as well as some coastal services. They were sometimes referred to as 'Yorkies' or 'Yorkshire Tanks', due to being manufactured in South Yorkshire. All thirty members of the class saw LMS and BR service and survived until 1952 when withdrawals started. The last two were withdrawn from service in 1960. All were scrapped.

No. 67464. LNER (NBR) 4-4-2 'Class C15' tank locomotive. Built in January 1913 by the Yorkshire Engine Company, Sheffield, for the North British Railway Company as NBR 'Class M' locomotive, number 6. Absorbed into the LNER in 1923 and re-classified as an LNER 'Class C15' tank engine. Later allocated LNER number 7464 followed by BR number 67464 in 1948. Withdrawn from service in August 1953 whilst based at Polmont Shed near Falkirk in Scotland. Scrapped at Kilmarnock Works shortly afterwards. (Photo P.65)

LNER 'Class V1' 2-6-2 tank locomotives: general information

A total of eighty-two LNER 'Class V1' 2-6-2, 3-cylinder tank locomotives, designed by Nigel Gresley, were built between 1930 and 1939. A further ten were built between 1939 and 1940 and although seemingly identical in appearance, they were fitted with higher pressure boilers and re-classified as LNER 'Class V3' engines. Both the 'Class V1' and 'Class V3' locomotives were originally intended for use as suburban passenger trains in the north London area and although one member of the class was successfully tested between Hitchin and London, the two classes were put to work in Scotland and the north east of England, where they remained.

Although primarily built for passenger use, they were classified as mixed traffic locomotives with a power classification of 4MT. A total of sixty-three of the 'Class V1' locomotives were rebuilt into 'Class V3' engines, taking their total to seventy-three. The remaining nineteen 'Class V1' engines were never rebuilt, but in 1953 their power classification was downgraded from 4MT to 3MT. The 'Class V3' locomotives retained their '4MT' status. Many of these tank engines saw service throughout the 1950s and the last 'Class V1' was withdrawn from service in 1962. The 'Class V3' became extinct in 1964 and sadly, none have been preserved.

No. 67629. LNER 2-6-2 'Class V1' tank locomotive. Built at Doncaster Works in February 1935 as LNER number 7629. Absorbed into BR in 1948 as BR number 67629. Withdrawn from service in May 1962 whilst working out of Parkhead Shed, Glasgow, and scrapped at Darlington Works in October 1962. (Photo P.86)

LNER (WD) 0-6-0 'Class J94' Austerity saddle tank locomotive: general information

Several hundred of these 0-6-0 saddle tank locomotives were built by various companies between 1945 and 1955. The locomotive was designed by Robert Arthur Riddles, former Assistant Chief Mechanical Engineer of the LMS Railway, who after the outbreak of war in 1939 had been appointed the Director of Transportation Equipment for the Ministry of Supply. The design was in response to the urgent need of powerful shunting engines for war use by the WD (War Department), engaged in the transportation of wartime goods and munitions. These austerity locomotives had a power classification of 4F and with their short wheelbase, they were ideal for use on the tight curves found in docks and harbours.

After the war, most of the locomotives were surplus to requirements, and sold to numerous industrial outlets such as railway companies, collieries, docks, and factories all over the country. The industrial demand for these engines was such that the building programme continued into 1954.

A total of seventy-five were sold to the LNER in 1946 and classified as LNER

'Class J94' locomotives. Many of these remarkable little engines survived until the last days of steam but just two BR examples were preserved after being withdrawn from service. However, fifty or so preserved from private sources make them numerically, the largest class of engine working on Britain's Heritage Railways.

No. 68068. LNER 0-6-0 'Class J94' saddle tank locomotive. Built in 1943 by the Hudswell Clarke Locomotive Company, Hunslet, Leeds, for the British Government (Ministry of Supply), for use by the War Department as shunting engine number WD 71475 during the Second World War. Sold to the LNER after the end of the conflict in 1946 and re-classified as LNER 'Class J94' locomotive number 8068 until 1948, when it was absorbed into BR as number 68068. Withdrawn from service in May 1965 whilst working out of Buxton Shed in Derbyshire. Scrapped by the Central Wagon Company, Ince, Wigan, in April 1966. (Photo P.87)

No. 68071. LNER 0-6-0 'Class J94' saddle tank locomotive. Built in June 1945 by Andrew Barclay Sons & Co., Kilmarnock, for the British Government (Ministry of Supply), for use by the War Department as shunting engine number WD 71532. Worked on the Longmoor Military Railway in Hampshire until she was purchased by the LNER in 1946. Re-classified as LNER 'Class J94' locomotive number 8071 until 1948, when she was absorbed into BR as number 68071. Withdrawn from service in August 1963, whilst based at Darlington Shed. Scrapped at Darlington Works later that month. (Photo P.87)

LNER 0-4-2T 'Class Z4 and Z5', GNSC 'Class X' and 'Class Y': general information

Just four 0-4-2 tank locomotives, designed by Thomas Heywood, were built for the GNSR in 1915 by the Manning Wardle Locomotive Company, Hunslet, Leeds. The four locomotives were built specifically to perform shunting duties at Aberdeen Docks, where they spent their entire working lives.

Two of the locomotives had 3ft 6in (1.07m) driving wheels and 13in x 20in (33cm x 51cm) cylinders and were classified as GNSR 'Class X' locomotives. The other two were slightly larger engines with 4ft (1.22m) driving wheels and 14in x 20in (35.6cm x 51cm) cylinders. They were classified as GNSR 'Class Y' locomotives.

The four engines were absorbed into LNER stock in 1923 and were re-classified. The 'Class X' locomotives became LNER 'Class Z4' engines and the 'Class Y' became LNER 'Class Z5' engines. The four locomotives all saw BR service in the 1950s. They were withdrawn from service between 1956 and 1960 before being scrapped.

No. 68190. LNER (GNSR) 0-4-2T 'Class Z4' tank locomotive. Built for the Great North of Scotland Railway by the Manning Wardle Locomotive Company, Leeds, in August 1915 as GNSR 'Class X' shunting engine number 114. Re-numbered GNSR number 43 later that year. Absorbed into the LNER in 1923 and re-classified as LNER 'Class Z4' number 6843. Allocated LNER number 8190 in 1946, followed by BR number 68190 in 1948. Withdrawn from service in April 1960 whilst based at Aberdeen Kittybrewster Shed. Scrapped at Inverurie Works the following month. Number 68190 spent her entire working life performing shunting duties at Aberdeen Docks. (Photo P.65)

No. 68192. LNER (GNSR) 0-4-2T 'Class Z5' tank locomotive. Built for the Great North of Scotland Railway by the Manning Wardle Locomotive Company, Leeds, in January 1915 as GNSR 'Class Y' locomotive number 116. Re-numbered GNSR number 30 later that year. Absorbed into the LNER in 1923 and re-classified as LNER 'Class Z5' number 6830. Allocated LNER number 8192 in 1946, followed by BR number 68192 in 1948. Withdrawn from service in April 1960 whilst based at Aberdeen Kittybrewster Shed. Scrapped at Inverurie Works the following month. (Photo P.65)

LNER 0-6-0T 'Class J63', GCR 'Class 5A' locomotive: general information

Just seven of these engines were designed by John Robinson and built between 1906 and 1914 for the GCR, to be used as shunting engines on Immingham Docks. They were based on the GCR 'Class 5' (LNER 'Class J62') saddle tank shunting engines but differed due to having side tanks as opposed to saddle tanks. Robinson classified them as GCR 'Class 5A' locomotives. They were referred to as 'Dock Tanks' and successfully used for many years at Immingham and Mersey Docks. They all survived into the BR era and were eventually withdrawn from service between 1953 and 1957.

No. 68204. LNER (GCR) 0-6-0T 'Class J63' tank locomotive. Built at Gorton Locomotive Works, Manchester, for the Great Central Railway in August 1906 as GCR 'Class 5A' number 60. Absorbed into the LNER in 1923 and re-classified as an LNER 'Class J63' engine. Allocated LNER number 5060 until 1946 when she was re-numbered 8204, followed by BR number 68204 in 1948. Withdrawn from service and scrapped in April 1956 whilst based at Immingham. This locomotive was the first member of the class to be built and worked her entire life as a shunting engine at Immingham Docks. (Photo P.25)

LNER 'Class J70' (Tram Engine), GER 'Class C53' locomotive: general information

Just twelve 0-6-0T 'Class C53' tram engines, designed by James Holden, were built at Stratford Works between 1903 and 1921. These peculiar looking engines resembled old railway guard's vans. Although classed as freight locomotives, primarily designed for shunting duties (power classification 0F – zero F), they were in fact 'Tram Engines', designed for use on the 'Wisbech & Upwell Tramway', a standard gauge light railway built by the GER in 1883. Part of the conditions imposed for operating this light railway was that for safety purposes, the tram engines had to be fitted with both cow-catchers and side shields covering the wheels. These tram engines inspired the Rev. W. Awdry to introduce 'Toby the Tram Engine' into his series of children's books.

All twelve of these engines were absorbed into the LNER in 1923 when they were re-classified as LNER 'Class J70' locomotives. One member of the class was scrapped in 1942 but the remainder saw BR service until they were withdrawn between 1949 and 1955. They were all scrapped.

No. 68222. LNER (GER) 0-6-0T 'Class J70' Tram locomotive. Built at Stratford Works in June 1914 for the Great Eastern Railway as a GER 'Class C53' Tram Engine number 128. Absorbed into the LNER in 1923 and re-classified as LNER 'Class J70' number 7128. Re-numbered LNER 8222 in 1946, followed by BR number 68222 in 1948. Withdrawn from service in February 1955 whilst based at Ipswich. Scrapped at Stratford Works in May 1955. (Photo P.25)

LNER 0-6-0T 'Class J73', NER 'Class L' locomotive: general information

A total of ten North Eastern Railway 'Class L' 0-6-0 tank locomotives, designed by Wilson Worsdell, were built at Gateshead Works between 1891 and 1892. They were his first ever locomotive design for the NER. They were built for use on the Redheugh and Quayside banks on either side of the River Tyne but were later allocated other duties. They could be seen in places that included Hull Docks, Gascoigne Marshalling Yard, Selby and West Hartlepool.

These 0-6-0 tank engines were very effective locomotives and as such their average working life span was an amazing sixty-five years. As well as working for the NER, they all saw service with the LNER ('Class J73' locomotives) and BR, before being withdrawn from service between 1955 and 1960. Sadly, none have been preserved.

No. 68359. LNER (NER) 0-6-0T 'Class J73' tank locomotive. Built at Gateshead NER Locomotive Works in April 1892, as NER 'Class L' locomotive number 548. Absorbed into the LNER in 1923 and re-classified as an LNER 'Class J73' locomotive. Allocated LNER number 8359, followed by BR number 68359 in 1948. Withdrawn from service in December 1959 whilst based at West Hartlepool Shed. Scrapped at Darlington Works in January 1960. (Photo P.24)

LNER 0-6-2T 'Class N1' (also GNR 'Class N1'), locomotive: general information

A total of fifty-six Great Northern Railway 'Class N1' 0-6-2 tank locomotives, designed by Henry Ivatt, were built at Doncaster Works between 1907 and 1912. They were designed as mixed traffic locomotives, to work in the north London area from locomotive sheds at King's Cross and Hornsey. Their power classification was 2 MT.

Fifty-two of these locomotives were fitted with condensing apparatus to enable them to work passenger trains through the London Metropolitan line tunnels to Moorgate. The remaining four were assigned to working in West Yorkshire. In the early 1920s, 'Class N1' locomotives began to be replaced by the more powerful 'Class N2' locomotives. The condensing apparatus was systematically removed from the engines which were then transferred to the Leeds, Bradford and Wakefield areas of West Yorkshire, where they continued working light freight and passenger services, as well as being used for station pilot duties. The whole class were withdrawn from service and scrapped between 1947 and 1959.

No. 69446. LNER (GNR) 0-6-2T 'Class N1' tank locomotive. Built at Doncaster Works in March 1910 as GNR 'Class N1' locomotive number 1566. Absorbed into the LNER in 1923 as an LNER 'Class N1' locomotive, number 4566. Re-numbered LNER 9446 in 1946, followed by BR number 69446 in 1948. Withdrawn from service in June 1953 whilst based at Leeds Copley Hill Shed. Scrapped at Doncaster Works later that month.

This engine displayed the number E 9446 in 1948. The numbers with prefix letters were applied for a three-month period only from January until March 1948 as a temporary measure until the allocation of all new BR numbers were finalised after nationalisation. The official BR numbers were then applied the next time the locomotive was in the works for re-painting. (Photo P.17)

No. 69452. LNER (GNR) 0-6-2T 'Class N1' tank locomotive. Built at Doncaster Works in November 1910 as GNR 'Class N1' locomotive number 1572. Absorbed into the LNER in 1923 as an LNER 'Class N1' locomotive, number 4572. Re-numbered LNER 9452 in 1946, followed by BR number 69452 in 1948. Withdrawn from service in March 1959 whilst based at Ardsley Shed in West Yorkshire. Scrapped at Doncaster Works in April 1959. (Photo P.17)

No. 69485. LNER (GNR) 0-6-2T 'Class N1' tank locomotive. Built at Doncaster Works in June 1912 as GNR 'Class N1' locomotive number 1605. Absorbed into the LNER in 1923 as an LNER 'Class N1' locomotive, number 4605. Re-numbered LNER 9485 in 1946, followed by BR number 69485 in 1948. Withdrawn from service in November 1954 whilst based at Bradford Hammerton Street Shed. Scrapped at Doncaster Works in December 1954. (Photo P.87)

LNER 0-6-2T 'Class N7', GER 'Class L77' tank locomotive: general information

A total of 134 Great Eastern Railway 'Class L77' 0-6-2 tank locomotives, designed by Alfred John Hill, were built for the GER and LNER between 1915 and 1928.

They were originally designed to work the busy suburban passenger services in and out of Liverpool Street in east London and some were fitted with condensing apparatus for working on the underground Metropolitan line (the apparatus was removed between 1935 and 1938).

In January 1923, when the GER was absorbed into the LNER, a total of twelve 'Class L77' locomotives had been built and a further ten were in the process of being built. The LNER re-classified the locomotives as LNER 'Class N7' engines, and Nigel Gresley decided to modify and improve the design and carry on building them for use as mixed traffic locomotives (power classification 3MT).

The building programme for LNER 'Class N7' locomotives ceased after 1928, although Gresley continued making improvements and modifications to the design until 1943. These improvements and modifications resulted in the class being sub-divided as follows:

- **N7.** GER 'Class L77' locomotives introduced in 1914, with Belpaire fireboxes.
- **N7/1.** LNER (Gresley) development of the GER 'Class L77'.
- **N7/2.** Locomotives built in 1926 with Belpaire fireboxes and fitted with long-travel valves.
- **N7/3.** Locomotives built in 1927 and rebuilds of existing engines with round-top fireboxes.
- **N7/4.** Introduced 1940. Some GER 'Class L77' engines rebuilt with round-top fireboxes.
- **N7/5.** Introduced 1943. Some N7/1 locomotives rebuilt with round-top fireboxes.

All the 'Class N7' locomotives were withdrawn from service between 1957 and 1962. Just one member of the class, number 69621, is preserved, at the East Anglia Railway Museum at Colchester in Essex.

No. 69692. LNER (GER) 0-6-2T 'Class N7/2' tank locomotive. Built by William Beardmore & Co., Glasgow in August 1927 as LNER number 2652. Re-numbered LNER 9692 in 1946, followed by BR number 69692 in 1948. Withdrawn from service in September 1962 whilst based at Stratford Shed in East London. Scrapped at Stratford Works in January 1963. (Photo P.63)

LNER 4-6-2T 'Class A5', GCR 'Class 9N' tank locomotive: general information

A total of forty-four, 'Class 9N' 4-6-2 ('Pacific') tank locomotives, designed by John Robinson, were built for the Great Central Railway and LNER between 1911 and 1926. They were designed for working express suburban passenger train services from London Marylebone to High Wycombe and Aylesbury, which they did for over thirty years. Their power classification was 4P (re-classified as 3P by BR in 1953).

Thirty-one of these 'Pacific' tank locomotives were built at Gorton Works, Manchester between 1911 and 1923 as GCR 'Class 9N' locomotives. After they were absorbed into the LNER they were re-classified as LNER 'Class A5' engines. Gresley then ordered an additional thirteen 'Class 5' engines, which were built for the LNER by Hawthorne Leslie in Newcastle, after which, the class was sub-divided as follows:

- **A5/1.** The original GCR 'Class 9N' engines built at Gorton became LNER 'Class A5/1' engines.
- **A5/2.** The additional LNER 'Class A5' engines built at Newcastle between 1923 and 1926 became LNER 'Class A5/2' engines.

'Class A5' locomotives could later be seen operating out of King's Cross and other parts of the country, including Bradford, Lincolnshire, Hull, Darlington and the North East.

Apart from one of the original 'Class 9N' locomotives which was scrapped in 1942, the whole class saw BR service until they were withdrawn for scrap between 1957 and 1961.

No. 69842. LNER (GCR) 4-6-2T 'Class A5/2' tank locomotive. Built at the Hawthorn Leslie Locomotive Works, Newcastle in March 1926, as LNER number 1790. Re-numbered 9842 in 1946, followed by BR number 69842 in 1948. Withdrawn from service in October 1958 whilst based at Thornaby Shed near Stockton-on-Tees. Stored at Darlington North Road Station for several months before being scrapped at Darlington Works in August 1959. She was the last member of the class to be built. (Photo P.86)

LNER 'Class A8' tank locomotive (Converted from Class H1): general information

A total of forty-five North Eastern Railway 'Class D' locomotives, designed by Vincent Raven, were built between 1913 and 1922 to work light express passenger trains, serving the towns and cities of Leeds, Harrogate, York, and Newcastle, as well as the coastal resorts of Bridlington, Scarborough and Whitby. These locomotives were 3-cylinder, 4-4-4 tank locomotives with a power classification of 4P (re-classified as 3P by BR in 1953). The locomotives were not very well liked by the crews, with their front double-bogie design causing excessive roll and an uncomfortable ride at high speeds.

The 'Class D' locomotives were absorbed into the LNER in 1923 and were re-classified as LNER 'Class H1' locomotives. Gresley decided to rebuild the whole class by removing the double-bogie design and introducing other modifications. As a result, all the 'Class H1' locomotives were converted from 4-4-4 tank engines into 4-6-2 tank engines between 1931 and 1936. The conversions resulted in a slight loss of speed but the locomotives became very stable and proved to be very successful and popular with the drivers. After receiving their conversions, they were re-classified as LNER 'Class A8' locomotives and the 'Class H1' engines became extinct.

The whole class saw BR service from 1948 and continued working until they were all withdrawn for scrap between 1957 and 1960.

No. 69860. LNER 4-6-2T 'Class A8' tank locomotive. Originally built at Darlington Works between 1913 and 1922 as an NER 'Class D' 4-4-4 tank locomotive. Absorbed into LNER in 1923 and re-classified as an LNER 'Class H1' locomotive number 2153. Rebuilt at Darlington in August 1934 as an LNER 'Class A8' 4-6-2 tank locomotive. Re-numbered LNER 9860 in 1946 (later BR number 69860). Withdrawn from service in June 1960 whilst based at Thornton Junction Shed, Fife, Scotland. Scrapped at Darlington Works in September 1960. (Photo P.14)

No. 69885. LNER 4-6-2T 'Class A8' tank locomotive. Originally built at Darlington Works between 1913 and 1922 as an NER 'Class D' 4-4-4 tank locomotive. Absorbed into LNER in 1923 and re-classified as an LNER 'Class H1' locomotive number 1526. Rebuilt at Darlington in May 1936 as an LNER 'Class A8' 4-6-2 tank locomotive. Re-numbered LNER 9885 in 1946 and BR 69885 from 1948. Withdrawn from service in June 1960 whilst based at Scarborough Shed. Scrapped at Darlington Works in September 1960. (Photo P.38)

LNER 4-8-0T 'Class T1', NER 'Class X' tank locomotive: general information

Vincent Raven built ten North Eastern Railway 'Class X' 4-8-0 tank locomotives between 1909 and 1910. They were originally designed by Wilson Worsdell for moving heavy freight and also for shunting duties in hump yards. These powerful 3-cylinder engines had a power classification of 7F to enable them to move heavy coal wagons. The ten were absorbed into the LNER in 1923 and re-classified as LNER 'Class T1' locomotives. Nigel Gresley built a further five 'Class T1' engines in 1925. Two of the original NER 'Class X' engines were scrapped in 1937 but the remainder survived into the BR era until they were withdrawn from service for scrap between 1955 and 1961.

No. 69918. LNER (NER) 4-8-0T 'Class T1' tank locomotive. Built at Darlington in November 1925 as LNER number 1656. Re-numbered LNER 9918 in 1946, followed by BR number 69918 in 1948. Withdrawn from service in October 1958 whilst based at Goole Shed. Scrapped at Darlington Works later that month. (Photo P.38)

LNER 0-8-0T 'Class Q1' tank locomotives (Rebuilds of 'Class Q4'): general information

A total of just thirteen 'Class Q1' tank engines were built for the LNER, having been converted from LNER 'Class Q4' tender locomotives.

A total of eighty-nine GCR, 'Class 8A' (LNER 'Class Q4'), 0-8-0 tender locomotives were built for the Great Central Railway between 1902 and 1911. They were designed by John Robinson to work heavy coal trains from the West Yorkshire Coalfields and over the Pennines into Lancashire. A large number of these engines were allocated to Mexborough Shed. The 'Class 8A' tender locomotives were absorbed into the LNER in 1923 and re-classified as LNER 'Class Q4' tender locomotives.

After the outbreak of war in 1939, the transportation of supplies and munitions began to take place on a large scale, creating the need for additional heavy shunting engines. In 1942, it was decided to convert twenty-five of the 'Class Q4' tender locomotives into 0-8-0 tank engines in order to meet these requirements.

Just thirteen 'Class Q4' locomotives were actually converted into tank engines between 1942 and 1945, after which wartime hostilities ceased and the rebuilding programme was abandoned.

The thirteen locomotives which had been rebuilt were re-classified as LNER 'Class Q1' tank locomotives. The first four to be converted in 1942/43 were fitted with 1,500 gallon side tanks and were later classified as 'Class Q1/1' locomotives, and the remaining nine, which were rebuilt between 1943 and 1945 with larger 2,000 gallon side tanks, were classified as 'Class Q1/2' locomotives.

No. 69933. LNER 0-8-0T 'Class Q1/2' tank locomotive. This locomotive started life as a GCR 'Class 8A', 0-8-0 tender engine number 87. She was absorbed into the LNER in 1923 and re-classified as an LNER 'Class Q4' locomotive, number 5087. In May 1944 she was rebuilt by the Kitson Locomotive Company in Leeds and converted into an LNER 'Class Q1' tank locomotive, retaining her LNER number. She was re-numbered LNER 9933 in 1946, and after nationalisation in 1948 she was allocated her BR number, 69933, working out of Gateshead Shed. She was later transferred to Selby Shed, from where she was withdrawn from service in December 1958. She was scrapped at Gorton Works, Manchester in January 1959. (Photo P.39)

LNER 2-8-8-2T 'Class U1' tank locomotive: general information

This LNER 'six-cylinder' locomotive was unique. She was the largest and most powerful locomotive ever built in Britain and was the sole member of the class (LNER 'Class U1'). She was never allocated a power classification and as such remained unclassified.

Number 69999 was designed for the LNER by Nigel Gresley and the Beyer, Peacock Company who built her. It was a Beyer-Garratt type locomotive, various designs of which had been successfully used for a number of years in many parts of the world. They were very popular locomotives in South Africa and South America.

Beyer-Garratt locomotives were in effect two engines in one, powered by one large boiler. In case of the 'Class U1', the 3-cylinder units used by Gresley were identical to those designed for his powerful 'Class O2' locomotives. For further general information about Beyer-Garratt type locomotives, please refer to the section of this book which covers

LMS, Fowler design, Beyer-Garratt locomotives (see pages 125 & 126).

The LNER 'Class U1' was specifically designed to be used as a banking engine, to assist heavy coal trains traversing the Worsborough incline in the heart of the South Yorkshire Coalfield, just outside Barnsley. The incline ran for seven miles, with a gradient of 1 in 40 stretching for three miles. It was the second steepest main line incline in Britain after the 1 in 37 Lickey incline in Worcestershire.

Number 69999 entered service in the summer of 1925 and was originally based in Barnsley, but was moved to Mexborough Shed shortly afterwards. From there, she successfully worked the incline for her entire LNER service. After the railways were nationalised in 1948, she entered the BR locomotive stock and continued working at Worsborough until the early 1950s, when she was put to use as a British Railways Engineers' Departmental Locomotive, to assist in the heavy engineering of a scheme which was carried out to electrify the main Manchester to Sheffield, Woodhead Railway Line. The scheme was fully completed in 1955.

After a short spell in storage at Gorton Locomotive Works, Manchester, number 69999 visited Bromsgrove in 1955 with a view to working as a banking engine on the famous Lickey incline, south of Birmingham. However, after a short trial period, it was decided that she was not suitable for that particular job and she returned to Gorton where she remained until being withdrawn from service in December 1955. She was scrapped at Doncaster Works in March 1956.

No. 69999. LNER 2-8-8-2T 'Class U1' (Gresley-Beyer-Garratt) tank locomotive. Built by Beyer, Peacock Ltd, Manchester in June 1925 as LNER 'Class U1' locomotive, number 2395. Renumbered LNER 9999 in 1946 followed by BR number 69999 in 1948. Withdrawn from service in December 1955 and scrapped at Doncaster Works in March 1956. (Photo P.91)

ABOUT THE AUTHOR

After leaving school, Malcolm Clegg enjoyed a thirty-year career with the British Transport Police. He served both in uniform and in CID, working mainly in South Wales, policing the railway network. He did, however, work for a number of years as a Docks Constable at Cardiff and Newport Docks and later worked for several years as a Uniform Sergeant at Swansea and Port Talbot Docks. In addition, almost a decade of his career was spent working at various locations in London.

The final ten years of his service were spent as a Detective Sergeant based in Swansea, investigating crimes committed on the Docks and Railway premises over an extensive area of South, and West Wales, which included Fishguard Harbour, incorporating the 'Sealink' passenger ferry services which operated at the time between Fishguard and Rosslare in Ireland.

After his retirement, he became an active member of the British Transport Police History Group (www.btphg.org.uk). He has carried out extensive research on behalf of the group and has written a number of articles. He is also the author of two books entitled THE LAST DAYS OF BRITISH STEAM and BRITISH STEAM LOCOMOTIVES BEFORE PRESERVATION. both published by Pen & Sword Books in 2020.